IFMBE Proceedings

Volume 79

Series Editor

Ratko Magjarevic, Faculty of Electrical Engineering and Computing, ZESOI, University of Zagreb, Zagreb, Croatia

Associate Editors

Piotr Ładyżyński, Warsaw, Poland

Fatimah Ibrahim, Department of Biomedical Engineering, Faculty of Engineering, University of Malaya, Kuala Lumpur, Malaysia

Igor Lackovic, Faculty of Electrical Engineering and Computing, University of Zagreb, Zagreb, Croatia

Emilio Sacristan Rock, Mexico DF, Mexico

The IFMBE Proceedings Book Series is an official publication of *the International Federation for Medical and Biological Engineering* (IFMBE). The series gathers the proceedings of various international conferences, which are either organized or endorsed by the Federation. Books published in this series report on cutting-edge findings and provide an informative survey on the most challenging topics and advances in the fields of medicine, biology, clinical engineering, and biophysics.

The series aims at disseminating high quality scientific information, encouraging both basic and applied research, and promoting world-wide collaboration between researchers and practitioners in the field of Medical and Biological Engineering.

Topics include, but are not limited to:

- Diagnostic Imaging, Image Processing, Biomedical Signal Processing
- Modeling and Simulation, Biomechanics
- Biomaterials, Cellular and Tissue Engineering
- Information and Communication in Medicine, Telemedicine and e-Health
- Instrumentation and Clinical Engineering
- Surgery, Minimal Invasive Interventions, Endoscopy and Image Guided Therapy
- Audiology, Ophthalmology, Emergency and Dental Medicine Applications
- Radiology, Radiation Oncology and Biological Effects of Radiation

IFMBE proceedings are indexed by by SCOPUS, EI Compendex, Japanese Science and Technology Agency (JST), SCImago.

Proposals can be submitted by contacting the Springer responsible editor shown on the series webpage (see "Contacts"), or by getting in touch with the series editor Ratko Magjarevic.

More information about this series at http://www.springer.com/series/7403

Chwee Teck Lim · Hwa Liang Leo ·
Raye Yeow
Editors

17th International Conference on Biomedical Engineering

Selected Contributions to ICBME-2019,
December 9–12, 2019, Singapore

Springer

Editors
Chwee Teck Lim
T-Lab
National University of Singapore
Singapore, Singapore

Hwa Liang Leo
National University of Singapore
Singapore, Singapore

Raye Yeow
Evolution Innovation Lab, Department
of Biomedical Engineering
National University of Singapore
Singapore, Singapore

ISSN 1680-0737   ISSN 1433-9277   (electronic)
IFMBE Proceedings
ISBN 978-3-030-62044-8   ISBN 978-3-030-62045-5   (eBook)
https://doi.org/10.1007/978-3-030-62045-5

This Springer imprint is published by the registered company Springer Nature Switzerland AG
The registered company address is: Gewerbestrasse 11, 6330 Cham, Switzerland

Organisation

Organising Committee

Chair

Chwee Teck Lim, National University of Singapore

Co-chair

Dean Ho, National University of Singapore

Past Chair

James Goh, National University of Singapore

Scientific Chair

Hwa Liang Leo, National University of Singapore

Scientific Co-chair

Raye Yeow, National University of Singapore

International Advisory Panel

Yoshinobu Baba, Nagoya University, Japan
Zhenan Bao, Stanford University, USA
Dominique Barthes-Biesel, University of Compiegne, France
Shu Chien, University of California, San Diego, USA
David Elad, Tel Aviv University, Israel
Yubo Fan, Beihang University, China
Gerhard Holzapfel, Graz University of Technology, Austria
Peter Hunter, University of Auckland, New Zealand
Roger Kamm, Massachusetts Institute of Technology, USA

Takehiko Kitamori, University of Tokyo, Japan
Geert Schmid-Schobein, University of California, San Diego, USA
Michael Shuler, Cornell University, USA
Fong-Chin Su, National Cheng Kung University, Taiwan
Mitsuo Umezu, Waseda University, Japan
David Weitz, Harvard University, USA

International Scientific Committee

Taiji Adachi, Kyoto University, Japan
Rashid Bashir, University of Illinois Urbana-Champaign, USA
Anthony Bull, Imperial College London, UK
Dennis Discher, University of Pennsylvania, USA
Cheng Dong, Penn State University, USA
Ross Ethier, Georgia Institute of Technology, USA
Amit Gefen, Tel Aviv University, Israel
Guy Genin, Washington University at St Louis, USA
Justin Gooding, University of New South Wales, Australia
Farshid Guilak, Washington University at St Louis, USA
X. Edward Guo, Columbia University, USA
Amy Herr, University of California, Berkeley, USA
Walter Herzog, University of Calgary, Canada
Marie-Christine Ho Ba Tho, University of Technology of Compiegne, France
I-Ming Hsing, Hong Kong University of Science and Technology, Hong Kong
Dietmar Hutmacher, Queensland University of Technology, Australia
Taeghwan Hyeon, Seoul National University, South Korea
Fatimah Ibrahim, University of Malaya, Malaysia
Takuji Ishikawa, Tohoku University, Japan
Timo Jämsä, University of Oulu, Finland
Shankar Krishnan, Wentworth University, USA
Ernesto Iadanza, University of Florence, Italy
Piotr Ladyzynski, Nalecz Institute of Biocybernetics and Biomedical Engineering, Poland
Abraham Lee, University of California, Irvine, USA
Peter Lee, University of Melbourne, Australia
Song Li, University of California, Los Angeles, USA
Feng-Huei Lin, National Taiwan University, Taiwan
Kang-Ping Lin, Chung Yuan Christian University, Taiwan
Mian Long, Institute of Mechanics, Chinese Academy of Sciences, China
Nigel Lovell, University of New South Wales, Australia
Ratko Magjarevic, University of Zagreb, Croatia
Josep Samitier Marti, Institute for Bioengineering of Catalonia, Spain
Takeo Matsumoto, Nagoya University, Japan

Leandro Pecchia, University of Warwick, UK
Beth Pruitt, University of California, Santa Barbara, USA
Heinz Redl, Ludwig Boltzmann Institute, Austria
John A. Rogers, Northwestern University, USA
Ichiro Sakuma, University of Tokyo, Japan
Vivek Sheenoy, University of Pennsylvania, USA
Masahiro Sokabe, Nagoya University, Japan
Yu Sun, University of Toronto, Canada
Shoji Takeuchi, University of Tokyo, Japan
Swee Hin Teoh, Nanyang Technological University, Singapore
Raymond Tong, Chinese University of Hong Kong, Hong Kong
Nitish Thakor, National University of Singapore, Singapore
Jaw-Lin Wang, National Taiwan University, Taiwan
Justin Cooper White, University of Queensland, Australia
Ed X. Wu, University of Hong Kong, Hong Kong
Younan Xia, Georgia Institute of Technology, USA
Guang-Zhong Yang, Imperial College London, UK
Leslie Yeo, RMIT University, Australia
Yong Ping Zheng, Hong Kong Polytechnic University, Hong Kong
Polona Žnidaršič Plazl, University of Ljubljana, Slovenia

Local Scientific Committee

Scientific Programme Chair

Hwa Liang Leo, National University of Singapore

Scientific Programme Co-chair

Raye Yeow, National University of Singapore

Members

Ali Asgar Bhagat, National University of Singapore
Chia-Hung Chen, National University of Singapore
Nanguang Chen, National University of Singapore
Peng Chen, Nanyang Technological University
Xiaodong Chen, Nanyang Technological University
Lih Feng Cheow, National University of Singapore
Sing Yian Chew, Nanyang Technological University
Keng Hwee Chiam, Institute of High Performance Computing, A*STAR
Desmond Chong, Singapore Institute of Technology
Mark Chong, Nanyang Technological University
Chee Kong Chui, National University of Singapore
Fangsen Cui, Institute of High Performance Computing, A*STAR

Eliza Fong, National University of Singapore
John Ho, National University of Singapore
Zhiwei Huang, National University of Singapore
James Chen Yong Kah, National University of Singapore
Sangho Kim, National University of Singapore
Vincent Lee, National University of Singapore
Jun Li, National University of Singapore
Yansong Miao, Nanyang Technological University
Kee Woei Ng, Nanyang Technological University
Kanyi Pu, Nanyang Technological University
Anqi Qiu, National University of Singapore
Hongliang Ren, National University of Singapore
Vinicius Rosa, National University Health System
Huilin Shao, National University of Singapore
Nguyen Vinh Tan, Institute of High-Performance Computing, A*STAR
Lay Poh Tan, Nanyang Technological University
Dalton Chor Yong Tay, Nanyang Technological University
Benjamin Tee, National University of Singapore
Yi-Chin Toh, National University of Singapore
Dieter Trau, National University of Singapore
Zhiping Wang, Singapore Institute of Manufacturing Technology, A*STAR
Wong Yoke Rung, Singapore General Hospital
Chen Jie Xu, Nanyang Technological University
Choon Hwai Yap, National University of Singapore
Ai Ye, Singapore University of Technology and Design
Joo Chuan Yeo, National University of Singapore
Haoyong Yu, National University of Singapore
Toyama Yusuke, National University of Singapore
Yong Zhang, National University of Singapore
Wenting Zhao, National University of Singapore
Liang Zhong, National Heart Centre

Themes and Topics

A: BioImaging and BioSignals

- Biomedical imaging
- Medical physics
- Biomedical instrumentation
- Biosignal processing

B: Bio-Micro/Nanotechnologies

- Biosensors
- BioMEMs
- Microfluidics and Lab-on-Chip
- Mutli-scale devices and systems
- Nanobiotechnology

C: Bio-Robotics and Medical Devices

- Computer-assisted surgery
- Medical robotics
- Rehabilitation engineering and assistive technology
- Exoskeletal and wearable devices
- Smart sensors
- Soft robotics

D: Biomaterials and Regenerative Medicine

- Artificial organs
- Biomaterials
- Controlled drug delivery
- Pharmaceutics
- Tissue engineering
- Bioprinting

E: BioMechanics and Mechanobiology

- Cardiovascular mechanics
- Cell and molecular mechanics
- Organ and tissue mechanics
- Orthopaedic biomechanics
- Mechanobiology
- Sports biomechanics and human performance

F: Engineering/Synthetic Biology

- Cell-free synthetic biology
- Engineering synthetic ecosystems
- Microbial production of drugs
- Artificial cells
- Synthetic viruses
- Genetic engineering
- Artificial photosynthesis
- Computational designs
- Tools and techniques

G: Neuroengineering/Neurotechnology

- Neuroimaging
- Brain–machine interfaces
- Advanced brain monitoring
- Innovations in neurotechnology

H: Big Data, Healthcare Analytics and AI

- Telemedicine and healthcare
- Data mining and machine learning
- Cybersecurity
- Bioinformatics and digital medicine
- Clinical decision support systems
- AI in healthcare
- Personalised medicine

I: Bioelectronics and Electroceuticals

- Bioelectronics devices and applications
- Emerging characterisation techniques, devices and materials
- Cell-material interface in natural systems

J: Computational Modeling

- Musculoskeletal
- Soft tissue
- Electro-physiology
- Biofluids

K: Clinical Engineering

- Health technology assessment
- Health technology Management
- Clinical engineering
- Human factors

L: Biomedical Engineering Education

- Disruptive technology for education
- Innovations in biomedical engineering education
- BioDesign in the BME curriculum

M: Human Disease Diagnosis and Therapy

- Diagnostics
- Therapeutics
- Immunotherapy
- Immuno-engineering
- Liquid biopsy
- Precision medicine

Preface

Jointly organised by the Department of Biomedical Engineering of the National University of Singapore (NUS) and the Biomedical Engineering Society (Singapore) (BES), the 17th International Conference on Biomedical Engineering (ICBME) 2019 was held at NUS from 9–12 December 2019. This conference aims to provide local and international participants an invaluable opportunity to stay current on the latest scientific developments and emerging challenges in the field of biomedical engineering.

ICBME 2019 congregated some 475 participants from 29 countries and received continued endorsement from the International Federation for Medical and Biological Engineering (IFMBE) as well as support from several local and regional societies. The event was also supported by the Santec Corporation, Polaris Science, Nature Springer and the Singapore Health Technologies Consortium. On the scientific programme front, we received close to 400 papers with more than 320 talks presented across a four-day programme. This book gathers a selection of 17 papers, selected upon a careful peer-review process by the scientific committees and conference reviewers.

On behalf of the Organising and Scientific Committees, we thank the International Advisory Board, International and Local Scientific Committees, speakers, participants, exhibitors, sponsors and supporting organisations for making ICBME 2019 a huge success. Our continued partnership is what helps to further advance the field of biomedical engineering.

Singapore

Chwee Teck Lim
Hwa Liang Leo
Raye Yeow

Contents

Pilot Study—Portable Evaporative Cooling System for Exercise-Induced Hyperthermia

Seng Sing Tan, Eng Koon Lim, and Chin Tiong Ng

1 Introduction

Every year, some 150 people fall prey to heat stroke in Singapore, where weather is hot and humid [10]. Most of them are young military personnel and athletes in long-distance runs. Mortality rates are directly related to the severity of hyperthermia as well as its duration [11]. Therefore, it is important to cool the body as quickly as possible for the greatest benefit, before heatstroke complications arise. Lowering the core body temperature to less than 40 °C within 30 min should be the primary aim of the treatment [2]. For immediate treatment on-site with limited portable equipment, the current treatment standard consists only of wet towels or cold packs applied over the groin and armpits to help bring the body temperature down in the interim before any effective cooling facilities become available. Heat stroke is a serious condition which can be fatal in the absence of prompt medical treatment. With proper treatment available on site, the chances of prolonging the patient's survival long enough for arrival of emergency services are boosted. An effective portable cooling system is essential for this purpose.

2 Background

Exercise-induced hyperthermia illness typically occurs over excessive hours of strenuous training in younger athletes or military recruits who participate in outdoor physical activity in hot and humid weather conditions, like in Singapore, for long

S. S. Tan (✉) · C. T. Ng
Nanyang Polytechnic, Singapore, Singapore
e-mail: tan_seng_sing@nyp.edu.sg

E. K. Lim
National Healthcare Group, Singapore, Singapore

© Springer Nature Switzerland AG 2021
C. T. Lim et al. (eds.), *17th International Conference on Biomedical Engineering*,
IFMBE Proceedings 79, https://doi.org/10.1007/978-3-030-62045-5_1

enough durations to cause the rate of heat production to exceed the capacity of the body to dissipate heat. When the body is unable to dissipate the extra heat, the core temperature will continue to rise and heat stroke occurs. In some cases, the sweating mechanism may also fail, resulting in an affected body that is not sweating. The most serious of these illnesses is exertional heat stroke (EHS), a condition marked by an elevated core body temperature greater than 40 °C and central nervous system dysfunction [9]. In more severe situations, exertional heat illnesses need to be assessed and appropriately treated to prevent potentially serious consequences or fatalities. A persistently high core body temperature above a critical threshold is what leads to serious damage to the body's functional systems. Trainers should, whenever possible, implement immediately the most effective cooling method. If cooling is delayed, the benefits decrease as a direct function of time.

Several cooling methods have been presented in the literature including immersion in ice water at different temperatures, evaporative cooling, ice pack application, pharmacological treatment and invasive techniques. Ice-water immersion has been claimed to provide the most efficient external cooling [2]. It can reduce the core body temperature by approximately 0.15 °C/min and is currently the gold standard recommended to treat athletes suffering from exertional heat stroke. However, using ice water to cool the body directly could have adverse effects, posing a risk of over-cooling hyperthermic individuals [3]. There are systems, such as cardiopulmonary bypass and recirculating coolants, which are better in monitoring and control of temperature drop, while reducing the core body temperature effectively within the patients. Although these can achieve a higher and more manageable cooling rate, they are extremely invasive and require a highly skilled team and hospital facilities to implement.

Given the time-critical nature of these life-threatening emergencies and the predominant occurrence of these injuries outside of hospitals, the time required to achieve cooling to their physiological levels became a critical parameter. Initiation of efficient cooling on site immediately after the EHS event is of crucial importance too [7]. The time to transport patients to a local emergency department is time lost that could be spent cooling them [12]. Delaying the cooling treatment, until more sophisticated equipment is available, may be harmful or even life-threatening. Heled [7] presented four cases of exertional heatstroke that differed in their severity and outcome. These cases demonstrated the importance of the "golden hour" as a life-saving time interval in the case of EHS. Even as simple as large quantities of tap water at the scenario site and during the evacuation has proven to achieve a cooling rate of 2.5–3.0 °C/h reduction in body temperature. These existing methods for induction of hypothermia are unfortunately either not effective enough or not portable. A fast effective response is required to save lives, particularly in remote areas. Thus, there remains a need for an effective and portable cooling system for patients with EHS.

3 Methodology

Current cooling methods for exercise-induced hyperthermia can be divided into two main categories: invasive devices reducing the core body temperature from within the patient and non-invasive devices cooling the body from outside. Current external cooling pads, such as Flex.Pad by EmCools, ArcticGel by Medivance, cooling blankets or ice bags, are much simpler for layperson use, relatively inexpensive and do not pose many of the risks associated with their more invasive internal counterparts, especially for patients treated outside the hospital environment. However, such approaches are slow in cooling. In general, the range of time required for these devices to achieve 4 °C of cooling spans from more than 2 h to almost 7 h [8]. Many studies [1, 5, 13, 14] have been conducted to understand the efficiency of evaporative cooling methods using the body cooling unit (BCU). A field-deployed BCU is currently used for cooling hyperthermia patients in most outdoor strenuous exercise, and this method of evaporative cooling has been demonstrated to be effective in the treatment of heat injuries [6]. Room-temperature water is used instead of cold water, as it eliminates the danger of cold-induced vasoconstriction and heat-producing shivering that could reduce the efficacy of the cooling measures. The cooling therapy will be terminated once the rectal temperature fell below 39 °C to avoid hypothermia. Evaluation of various non-invasive cooling systems was conducted [14] and suggested that ThermoSuit offered the greatest mean absolute cooling rate. It was similar to ice-water immersion, except that it needed a pumping system to circulate cold water at a controlled flow rate and temperature. Although the suit was claimed to be portable, it required a power supply. EmCools Flex.Pad was more portable and offered the next best cooling rate. BCU has no clinical difference from EmCools pads but BCU was also not as well portable.

In this pilot study, we present a portable evaporative cooling system (PECS) which makes use of water sprinkled over the patient and a continuous flow of dehumidified air to stimulate the evaporation effect and promote an effective whole-body cooling through phase-change. Air flow with lower relative humidity can stimulate the evaporative effect. Thus, the water on the patient's wet body will be cooler during the evaporation process. Based on the principles of psychrometry, the evaporating rate of a wet surface could be moderated by the flow rate of the circulating air over the surface, consequently lowering the surface temperature. This approach could achieve a cooling temperature approach to wet-bulb temperature which is lower than the air flow temperature. Like the cooling tower, heat energy removed from water through the evaporating process could result a significant decrease in the water temperature, even below the ambient air temperature. Besides the airflow rate, the humidity of the supply air also plays an important role in optimizing the evaporating cooling effect. If the circulating air supply has a lower relative humidity, more water could be evaporated into the circulating air. With a higher flow rate of supply air, more heat will be removed through the evaporation process. Both the air supply humidity and velocity affect the evaporation rate of the water which is sprinkled out through nozzles and lands on the patient below, wetting the whole body and removing heat

Supply of pressurised air with low relative humidity through flexible tubing.

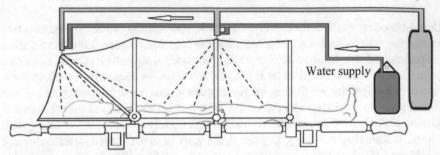

Water supply

Fig. 1 Schematic setup of portable evaporative cooling system

from it. A model was developed to conduct a pilot study to evaluate the effectiveness of a portable evaporative cooling system integrated on a MedEvac4 tactical stretcher, as shown in Fig. 1.

A retractable flexible framework lined with nozzles was fixed over the stretcher to enclose the patient. Flexible tubing supplying high velocity airflow was connected to these nozzles from a refillable compressed air cylinder. Water was supplied by separate tubing connected from a small bag of water supply near to the tip of each nozzle. The high velocity flow of air created a low static pressure at the exit of the nozzles, siphoning water from the water supply bag and sprinkling it out through the nozzles over a heat-stroke patient. This negates the need of an electrical water pump and thus makes the system portable. In this study, air supply was configured to flow at a rate of 260 L/min continuously, while the water supply was controlled to flow at 30 mL/min and stop flowing alternately at 3 min intervals. Without the need to immerse the whole body of the patient in water, a monitoring system could be connected to continue assessing the vital signs of the patient, allowing for close monitoring of the rate of cooling and adjustment of the cooling therapy if needed. This pilot study was designed to use the smallest number of animals possible to adhere with the requirements of the National Advisory Committee for Laboratory Animal Research, Singapore (NACLAR) guidelines, while adhering to the objectives of this study. The porcine model was chosen because of its similarity to humans in terms of anatomy and size. Two healthy animals were thus used in the trials, one for the test article, the portable evaporative cooling system (PECS), and another for a typical cooling pad (EmCools Flex.Pad) as a control article to compare their cooling rates and evaluate the effectiveness of PECS.

4 Preparation and Setup

The two porcines used in this study were acclimatized for 14 days prior to experimental phase and the study animals were experimentally naïve. The animals were appropriately evaluated and examined to ensure that they were in good clinical condition. Each animal was shaved at the front, back, and facial regions, as its hair being much longer than human could deviate the outcomes from the trials. Electrocardiogram (ECG), respiration rate, heart rate and oxygen saturation were monitored continuously during the procedures. The EmCools pads were stored at -20 °C for approximately 24 h prior to usage. Those pads that were not ready for use, would have beige or black color indicators on them. When the pad indicators turned blue, they were ready for use.

On the day of every trial, the animal was pre-medicated using atropine sulphate (50 μg/kg, IM). After which, it was anesthetized using TKX (0.05 mL/kg, IM). Anesthesia was maintained using 1–3% isoflurane. The theatre room temperature was regulated at 25 °C (± 2 °C) during the trials. A temperature probe was inserted into the esophagus and marked to establish the appropriate length of probe being inserted.

5 Experiment

The objective of the animal study was to evaluate the performance of the test article, portable evaporative cooling system (PECS) as compared to the control article, EmCools pads on a porcine heat-stroke model. To be considered effective, the systems should be able to reduce hyperthermic body temperature to 38 °C or back to physiological levels within 30 min when used as intended. In each trial, the animal was first warmed up to a target temperature of 41 °C before starting the cooling treatment. The core temperature was then monitored and recorded to analyse the cooling rate until the temperature reached 38 °C. The animal was recovered after each trial. In Study #1, the test and the control articles were each trialed on two different animals. After seven days, the trials were repeated in Study #2 on the same two animals from Study #1.

Warming Phase—Electrical warming blankets were used to warm the animals, which were first wrapped with cotton blankets to prevent direct burns. The animals' ears were covered well to prevent heat dissipation through the ears. 0.9% saline at 39 °C was also prepared for IV infusion to warm the animal via slow drip. The animals' core temperature, heart rate, and room temperature readings were recorded every 5–10 min to avoid overheating beyond the targeted temperature of 41 °C, and to observe for any unstable vital signs during procedural monitoring.

Cooling phase—The warm saline IV infusion and blankets were first removed from the hyperthermia-induced animal before commencing the cooling phase. For the control trials, six EmCools pads were pasted onto the animal's body (Chest, abdomen, back, extremities). Each pad was confirmed to be ready for use by the blue colour of their indicators before use in the trials. The pads were flexed horizontally and vertically to crack the channels between the cooling cells. The protective foil was then peeled from the back of each pad and immediately applied directly onto the body's surface. Each cooling pad was pressed down for 3–5 s to ensure full adherence of the adhesive layer. Animal core temperature, heart rate, and room temperature were recorded every minute after cooling pad application. As soon as the target temperature of 38 °C was reached, the pads were carefully removed to stop the cooling treatment.

For the test article trials, another hyperthermia-induced animal was transferred onto the MedEvac4 tactical stretcher and then enclosed by the PECS which was attached onto the stretcher. Animal's heart rate and room temperature were recorded every minute once the supply of compressed air cylinder was turned on, while the timing was recorded whenever the core temperature decreased by every 0.1 °C. The cooling treatment was continued until the body temperature reached the target 38 °C or until 30 min had passed, whichever earlier.

At the end of the cooling phase, the animals were then recovered from the treatments. These procedures were repeated after seven days for both the test and control articles with their respective animals. Cooling temperature profiles were plotted and compared.

6 Results and Analysis

This animal study showed that it was possible to use the Portable Evaporative Cooling System (PECS) to reduce body temperature back to physiological levels within 30 min as intended. The system has achieved the recommended cooling rate of 0.15 °C/min which is currently the recommended to treat athletes suffering from exertional heat stroke [2].

In all trials, the animals were warmed with the same set of procedures during the warming phase. The targeted end warming phase temperature was set at 41 °C. However, it took time to transfer the animals from the warming setup to the cooling system. The time taken to shift the animals to the stretcher and set up the system before initiating the cooling process varied slightly. Inevitably, the temperature probe was shifted in the esophagus during the transferring process. When the probe was repositioned through the throat back into the esophagus to the marked position, the recorded initial temperatures were slightly different for all the trials when the cooling stages commenced.

In Study #1, the first trial was to determine the time taken to cool down the hyperthermia-induced animal using EmCools Flex.Pad as a control article. During the warming phase, the animal became tachycardia. Its heart rate proceeded to increase quite significantly while its core temperature was approaching to the target temperature of 41 °C. In order to ensure the safety of the animal, the heating phase was terminated before the targeted temperature could be achieved as planned. By the time the cooling phase commenced with the EmCools pads, the animal's core body temperature was measured at only 40.6 °C. The time taken to bring the temperature down to 38 °C was short, with an average cooling rate of 0.22 °C/min.

In the other trial, the objective was to determine the time taken to cool down another hyperthermia-induced animal using PECS. During the warming phase, the

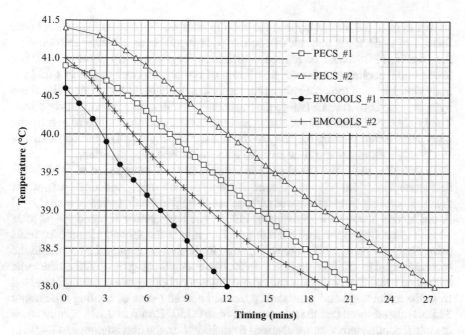

Fig. 2 Cooling profile comparison between EmCools and PECS

animal could be warmed to slightly above the desired initial target temperature of
41 °C as it did not become tachycardic with the dangerously high heart rate, unlike in
the first trial. After settling the animal onto the stretcher enclosed with the PECS, the
cooling phase started at an initial temperature of 40.9 °C. The cooling temperature
profile was plotted and analysed. It showed a lower cooling rate at the beginning,
which increased gradually till it reached at a steady cooling rate of 0.15 °C/min after
3 min as shown in Fig. 2.

Another two trials were conducted on the same two animals after 7 days for Study
#2. By the time the EmCools pads were pasted across the body of the animal, the
initial cooling temperature was recorded as 41.0 °C. The temperature profile showed
that it was able to cool the body faster initially. Cooling of the body from 41 to 38 °C
was achieved in less than 19.5 min using the EmCools pads, with an average cooling
rate of 0.15 °C/min in Study #2.

For the trial using PECS in Study #2, the cooling process started at 41.4 °C.
Similarly, the initial cooling rate was slower during the initial 3 min. Thereafter, the
cooling of the body was able to be maintained at a steady rate of 0.136 °C/min. All
animals were fully recovered from anesthesia at the end of all trials.

Using EmCools pads could cool down the body of the heatstroke patient rapidly at
the initial cooling. It has the advantage of cooling down the heat stroke patient faster
only for the first 12 min as the pads have direct contact to the body. However, as it
continued beyond 12 min, the rate of cooling turned slower, prompting the operator
to replace with a new one if the rapid cooling is essential. As the EmCools pads

started to melt, the melted layer in the cooling pads would form an insulation layer. Instead of replacing a new set of cooling pads regularly to achieve rapid cooling for 30 min, PECS could be a good substitution for a consistent cooling rate. The lower initial cooling rate for the first 3 min of cooling was due to the fact that the body was dry and it took sometimes for the system to disperse the water and wet the hyperthermic body surface before the evaporative cooling process could take effect. The PECS cooling rate was consistent and stable after the initial stage of getting the body surface wet. Wetting the body of the heatstroke patient at the start before using PECS might eliminate the slow onset to achieve the best cooling effect.

In this pilot study, the initial cooling temperature was recorded at 40.6 °C in the first attempt and 41.0 °C in the second attempt for trials using EmCools pads, while the initial cooling temperature was recorded at 40.9 °C in the first attempt and 41.4 °C in the second attempt for trials using PECS. The temperature reached 38 °C after 12 min of cooling in the first attempt and 19.5 min in the second attempt for trials using EmCools pads, while it took 21.5 min in the first attempt and 27.5 min in the second attempt for trials using PECS. In the two attempts using the PECS on the same animal on two different days, results showed that the cooling rate regularly increased from the initial stage and consistently reached a steady state of cooling after about 3 min. It also showed that the rate reduced from 0.150 °C/min to 0.136 °C/min when the initial cooling temperature changed from 40.9 °C in the first attempt to 41.4 °C in the second attempt. With different initial cooling temperatures, the results of all four trials showed that the higher the initial temperature at the beginning of the cooling process, the slower the cooling rate was. From this, it could be deduced that a patient could experience a lower cooling rate if he is more severely hyperthermic, regardless of the cooling system used.

7 Discussion

Currently, it is suggested that ice-water immersion could provide the most efficient cooling [9] and is the gold standard recommended [2] to treat athletes suffering exertional heat stroke. However, trying to cool a person suffering from heat stroke in ice-cold water directly is not highly recommended [10] because it causes the patient to shiver—the natural instinct of the body to keep warm—and that causes heat to be retained instead. Some experts also suggested sudden ice-water immersion might actually cause vasoconstriction [11] and make it harder for heat to dissipate from the body. Besides such counter-productive effects, there are the difficulties of handling other electronic monitoring systems during the cooling process. The use of ice-water immersion might pose a risk of overcooling hyperthermic individuals as the result of a lack of rectal temperature monitoring. Another existing evaporative cooling system includes a fan for generating a path of air flow and dispersing water in the path of air flow [4]. It creates a cool microclimate at the skin by wetting the skin and directing a flow of air over the patient to evaporate water for the cooling effect. Instead of using ice water to drench the body, such an external evaporative convective cooling is more easily handled together with other electronic monitoring devices. However, this system is normally set up within the hospital and may not be as portable as our PECS. Patients should always be cooled first at the scene to minimise heat damage and maximise survival outcomes. Furthermore, most of these evaporative cooling systems use electrical fans to direct ambient air to patients. Especially in a hot and humid city, like Singapore, such systems would be less effective without consideration of lowering the relative humidity of the circulating air. EmCools pads were recommended [14] for its comparable cooling rate. It also has the advantage of portability.

The current PECS offers external cooling which is much simpler to implement, relatively inexpensive and also suitable for deploying on-site. Using compressed air cylinder not only making it more portable as compared to the other evaporative cooling systems which use electrical fans, the air supply would have lower relative humidity for better evaporation cooling effect. It does not pose many of the risks associated with more invasive internal cooling systems, especially for patients outside the hospital setting. The ultimate goal is to bring the core body temperature down to less than 39 °C within 30 min at a rate not less than 0.15 °C/min even before transporting to hospital. This pilot study addressed the issue on the immediate treatment needed on site and its effectiveness after careful consideration on the relative humidity and velocity of the air supply to optimize the evaporative cooling effect. Although it is not as portable as EmCools pads, it has a steady cooling rate for a longer duration. However, large compressed air cylinder is required to provide substantial supply of air. If the option of using liquidized gas could be further explored in the subsequent study, the system could further improve its portability and treatment duration.

8 Conclusion

Several existing cooling methods for induction of hypothermia are either less effective or not portable. Thus, it is essential to develop an effective and portable system available for deployment at the scene. This pilot study demonstrated that a working prototype of an effective portable evaporative cooling system (PECS) could achieve the desired baseline cooling rate of 0.15 °C/min. It comprised an automated-control system to disperse water particles to wet the surface of the patient's body, and a continuous supply of dry air to cool the wet surface and promote the evaporative cooling effect. Our findings showed that the EmCools Flex.pad had the advantage of cooling down the heat stroke patient faster only for the first 12 min, as the pads have direct contact to the body. The cooling rate thereafter was slowed down by the insulative melted gel layer. The proposed evaporative cooling system would be a good substitution for treatment longer than 12 min. Such finding is a particularly important since at least 30 min of cooling is required for exertional heat stroke patients. While the PECS' initial 3 min of cooling was not as effective, PECS' cooling rate became consistent and stable beyond the initial transient stage of getting the body surface of the heatstroke patient wet. We suggest wetting the body surface with splash of water first before using PECS to achieve the best cooling effect from the start. More experimental investigations should be studied further with different flow rates and intervals to optimize the cooling effectiveness and ensure consistency of the results.

References

1. Bouchama, A., Dehbi, M., Chaves-Carballo, E.: Cooling and hemodynamic management in heatstroke: practical recommendations. Crit Care. **11**(3), R54 (2007)
2. Casa, D.J., McDermott, B.P., Lee, E.C., Yeargin, S.W., Armstrong, L.E., Maresh, C.M.: Cold water immersion: the gold standard for exertional heatstroke treatment. Exerc Sport Sci Rev **35**(3), 141–149 (2007)
3. Gagnon, D., Lemire, B.B., Casa, D.J., Kenny, G.P.: Cold-water immersion and the treatment of hyperthermia: using 38.6 °C as a safe rectal temperature cooling limit. J. Athl. Train. **45**(5), 439–444 (2010)
4. Gordon, LR.: Droplet injection system for evaporative cooling of animals. US patent US4693852 (1987)
5. Harker, J., Gibson, P.: Heat-stroke: a review of rapid cooling techniques. Intens. Crit. Care Nurs. **11**(4), 198–202 (1995)
6. Hee-Nee, P., Rupeng, M., Lee, V.J., Chua, W.C., Seet, B.: Treatment of exertional heat injuries with portable body cooling unit in a mass endurance event. Am. J. Emerg. Med. **28**(2), 246–248 (2010)
7. Heled, Y., Rav-Acha, M., Shani, Y., Epstein, Y., Moran, D.S.: The "Golden hour" for heatstroke treatmnent. Mil Med. **169**, 184–186 (2004)
8. Lampe, J.W., Becker, L.B.: Rapid cooling for saving lives: a bioengineering opportunity. Expert Rev. Med. Dev. **4**(4), 441–446 (2007)
9. McDermott, B.P., Casa, D.J., Ganio, M.S., et al.: Acute whole-body cooling for exercise-induced hyperthermia: a systematic review. J. Athl. Train. **44**(1), 84–93 (2009)
10. Ng, W.C.: Keep Your Cool. *The Straits Times.* 23 Aug, 2012; Mind Your Body
11. Raukar, N., Lemieus, R.S., Casa, D.J., Katch, R.K.: Identification and treatment of exertional heat stroke in the prehospital setting. J. Emerg. Med. Serv. **42**(5)
12. Sloan, B.K., Kraft, E.M., Clark, D., Schmeissing, S.W., Byrne, B.C., Rusyniak, D.E.: On-site treatment of exertional heat stroke. Am. J. Sports Med. **43**(4), 823–829 (2015)
13. Smith, J.E.: Cooling methods used in the treatment of exertional heat illness. Br. J. Sports Med. **39**(8), 503–507 (2005)
14. Tan, P.M.S., Teo, E.Y.N., Ali, N.B., et al.: Evaluation of various cooling systems after exercise-induced hyperthermia. J. Athl. Train. **52**(2), 108–116 (2017)

Meshless Method for Numerical Solution of Fractional Pennes Bioheat Equation

Hitesh Bansu and Sushil Kumar

1 Introduction

Fractional calculus and partial differential equations of fractional order have established a new research trend because of their ability to provide in-depth and accurate analysis of the model [3]. In modern clinical treatments like cryosurgery, cancer hyperthermia etc. behaviour of heat diffusion in living tissue is an interesting subject [5, 6]. Hence, some governing equations have been formulated to analyse the thermal behaviour of the biological tissues. Among these proposed models, the Pennes equation is the most extensively used due to its effectiveness [2, 4].

In the present study, we have considered the Pennes bioheat model as [8]

$$\rho_t c_t \frac{\partial T}{\partial t} = k \frac{\partial^2 T}{\partial x^2} + W_b c_b (T_a - T) + q_{met}, \quad x \in (0, L), \ t > 0, \tag{1}$$

with initial and boundary conditions

$$T(x, 0) = T_a$$

$$k \frac{\partial T}{\partial x}\bigg|_{x=0} = q_0 \quad k \frac{\partial T}{\partial x}\bigg|_{x=L} = 0 \tag{2}$$

where $\rho, c, T, k, t, x, T_a, W_b$ and q_{met} are density, specific heat, temperature, thermal conductivity, time, distance, artillery temperature, blood perfusion rate, and metabolic heat respectively.

H. Bansu (✉) · S. Kumar
S. V. National Institute of Technology, Surat, India
e-mail: hiteshbansu@gmail.com

S. Kumar
e-mail: sushilk@amhd.svnit.ac.in

© Springer Nature Switzerland AG 2021
C. T. Lim et al. (eds.), *17th International Conference on Biomedical Engineering*,
IFMBE Proceedings 79, https://doi.org/10.1007/978-3-030-62045-5_2

Due to its importance in clinical research, various techniques have been developed to solve the Penns bioheat equation. Further, while dealing with fractional PDEs numerical methods are always a priority. Ferras et al. [6] applied implicit finite difference scheme to solve space FPBE and the same method has been used by Damor et al. [4] to go through the parametric study of time FPBE with sinusoidal heat flux. Damor et al. [2] derived numerical solution of time FPBE with constant and transient heat flux.

In our work, we have solved the FPBE by the approach of collocation using Chebyshev polynomials and radial basis functions (RBFs). This approach is meshfree and implementable with the simultaneous fractional orders.

2 Preliminaries

Here we have discussed preliminaries and definition of fractional derivative, Chebyshev polynomials, and RBFs.

2.1 Fractional Derivatives

Definition 1 The left and right Caputo fractional derivative of order α of a function $f(x)$ for $\alpha > 0$, $(\alpha \in \mathbb{R})$ is [9]

$$\left(^{C}D_{a+}^{\alpha}f\right)(t) = \frac{1}{\Gamma(n-\alpha)} \int_{a}^{t} \frac{f^{(n)}(s)\,ds}{(t-s)^{1+\alpha-n}}$$

$$\left(^{C}D_{b-}^{\alpha}f\right)(t) = \frac{(-1)^{n}}{\Gamma(n-\alpha)} \int_{t}^{b} \frac{f^{(n)}(s)\,ds}{(s-t)^{1+\alpha-n}}$$

Definition 2 The Caputo fractional derivative of power of a function x^{p}, $p \geq 0$ is [9]

$$D^{\alpha}x^{p} = \begin{cases} \frac{\Gamma(p+1)}{\Gamma(p-\alpha+1)}x^{p-\alpha}, & \text{for } p \geq \lceil \alpha \rceil \\ 0, & \text{for } p < \lceil \alpha \rceil \end{cases}$$

2.2 Chebyshev Polynomials

The Chebyshev polynomials $T_m(x)$ have the analytic form as [1, 7]

$$T_m(x) = \sum_{l=0}^{\lfloor m/2 \rfloor} (-1)^l 2^{m-2l-1} \frac{m(m-l-1)!}{l!(m-2l)!} x^{m-2l}$$

where integer part of $m/2$ is indicated by $\lfloor m/2 \rfloor$. Now to use these polynomials on $[0, t]$ its necessary to perform the change of variable $s = \frac{2x}{t} - 1$, $s \in [-1, 1]$. Hence the shifted Chebyshev polynomials can be defined as $T_m^*(x) = T_m\left(\frac{2x}{t} - 1\right)$ and will have the analytic form as [1, 7]

$$T_m^*(x) = \sum_{l=0}^{m} (-1)^{m-l} 2^{2l} \frac{m(m+l-1)!}{(2l)!(m-l)!t^l} x^l.$$

The shifted Chebyshev polynomial will have the Caputo fractional derivative $D^\beta T_m^*(x)$ as

$$D^\beta T_m^*(x) = \sum_{l=\lceil \beta \rceil}^{m} h_{l,m,\beta} x^{l-\beta}, \quad m \geq \lceil \beta \rceil$$

where

$$h_{l,m,\beta} = (-1)^{m-l} 2^{2l} \frac{m(m+l-1)!}{(2l)!(m-l)!t^l} \frac{\Gamma(l+1)}{\Gamma(l+1-\beta)}.$$

2.3 Radial Basis Functions

In the last few decades, RBFs have become a successful tool for scattered data interpolation in which the function $f(x)$ can be estimated as [1]

$$f(x) = \sum_{i=1}^{N} \lambda_i \phi_i(r),$$

where number of data points is indicated by N. The coefficients $\{\lambda_i\}_{i=1}^{N}$ are to be investigated. $\phi(r)$ is any RBF and the Euclidean distance from point x to x_i is denoted by $r = \|x - x_i\|$.

Some commonly used RBFs are the multiquadric $\phi(r) = (r^2 + \epsilon^2)^{\alpha/2}$, $\alpha = -1, 1, 3, 5, \ldots$, the Gaussian $\phi(r) = e^{-(\epsilon r)^2}$, the conical type $\phi(r) = r^m$, $m = 1, 3, 5, \ldots$ and the polyharmonic splines $\phi(r) = r^m \log r$, $m = 2, 4, 6, \ldots$. In this study the implemented RBF is conical type with $n = 3$.

3   Heat Transfer Model

Penns bioheat model is commonly used to examine the heat transfer in living tissue [8]. Here we have obtained space and time FPBE by replacing space and time derivative with Caputo fractional derivative.

$$\rho_t c_t \frac{\partial^\alpha T}{\partial t^\alpha} = k \frac{\partial^\beta T}{\partial x^\beta} + W_b c_b \left(T_a - T\right) + q_{met}, \quad 0 < \alpha \le 1, \quad 1 < \beta \le 2. \tag{3}$$

For $\alpha = 1$ and $\beta = 2$ Eq. (3) coincides with classical Penns bioheat equation as in (1).

4   Numerical Solution

For the solution of bioheat model as given in (1)–(2), the function $T(x, t)$ can be approximated in terms of RBFs and Chebyshev polynomials as [1]

$$T(x, t) \approx \sum_{i=1}^{N} \sum_{j=1}^{n} \mathbb{T}_j(t)\, c_{ji} \Phi_i(x)$$

$$= \mathbb{T}(t)\, C\, \Phi(x) \tag{4}$$

where $\Phi(x) = [\phi_1(x), \phi_2(x), \phi_3(x), \ldots, \phi_N(x)]^T$, $\mathbb{T}(t) = \big[T_1(t), T_2(t), T_3(t), \ldots,$ $T_n(x)\big]$ are cubic RBFs and Chebyshev polynomials respectively. N and n (both $\in N$) are discretization parameters for space and time respectively. The form of unknowns c_{ji}s is

$$C = \begin{bmatrix} c_{11} & c_{12} & \cdots & c_{1N} \\ c_{21} & c_{22} & \cdots & c_{2N} \\ \vdots & \vdots & \ddots & \vdots \\ c_{n1} & c_{n2} & \cdots & c_{nN} \end{bmatrix}.$$

For discretization of space and time we have used uniform nodes on $[c, d]$ as

$$z_m = z_{m-1} + \frac{d - c}{p - 1}; \quad m = 1, 2, \ldots, p, \quad z_0 = c.$$

Here discretization for space and time are not depending on each other.

Using Eq. (4) its easy to derive

$$\,_0^C D_t{}^\alpha T\,(x,t) = \,_0^C D_t{}^\alpha \,(\mathbb{T}\,C\,\Phi) = \left[\,^C D_t{}^\alpha \mathbb{T}\right] C\Phi = \mathbb{T}^\alpha\,C\,\Phi, \tag{5}$$

$$\,_0^C D_t{}^\beta T\,(x,t) = \,_0^C D_t{}^\beta \,(\mathbb{T}\,C\,\Phi) = \mathbb{T}\,C\left[\,^C D_t{}^\beta \Phi\right] = \mathbb{T}\,C\,\Phi^\beta. \tag{6}$$

Putting Eqs. (5) and (6) in Eq. (3)

$$\rho_t c_t \left\{\mathbb{T}^\alpha\,C\,\Phi\right\} = k\left\{\mathbb{T}\,C\,\Phi^\beta\right\} + W_b c_b \left\{T_a - \{\mathbb{T}\,C\,\Phi\}\right\} + q_{met}. \tag{7}$$

Collocating (7) in $N-2$ and $n-1$ uniform nodes will give $(N-2)(n-1)$ equations as

$$\rho_t c_t \{M_1\,C\,L\} = k\{M\,C\,L_1\} + W_b c_b Ta - W_b c_b \{M\,C\,L\} + q_{met} \tag{8}$$

where

$$L = \begin{bmatrix} \phi_1\,(x_2) & \phi_2\,(x_2) & \cdots & \phi_N\,(x_2) \\ \phi_1\,(x_3) & \phi_2\,(x_3) & \cdots & \phi_N\,(x_3) \\ \vdots & \vdots & \cdots & \vdots \\ \phi_1\,(x_{N-1}) & \phi_2\,(x_{N-1}) & \cdots & \phi_N\,(x_{N-1}) \end{bmatrix}$$

$$M_1 = \begin{bmatrix} \,_0^C D_x^\alpha T_1\,(t_2) & \,_0^C D_x^\alpha T_2\,(t_2) & \cdots & \,_0^C D_x^\alpha T_n\,(t_2) \\ \,_0^C D_x^\alpha T_1\,(t_3) & \,_0^C D_x^\alpha T_2\,(t_3) & \cdots & \,_0^C D_x^\alpha T_n\,(t_3) \\ \vdots & \vdots & \cdots & \vdots \\ \,_0^C D_x^\alpha T_1\,(t_n) & \,_0^C D_x^\alpha T_2\,(t_n) & \cdots & \,_0^C D_x^\alpha T_n\,(t_n) \end{bmatrix}$$

$$M = \begin{bmatrix} T_1\,(t_2) & T_2\,(t_2) & \cdots & T_n\,(t_2) \\ T_1\,(t_3) & T_2\,(t_3) & \cdots & T_n\,(t_3) \\ \vdots & \vdots & \cdots & \vdots \\ T_1\,(t_n) & T_2\,(t_n) & \cdots & T_n\,(t_n) \end{bmatrix}$$

$$L_1 = \begin{bmatrix} \,_0^C D_x^\beta \phi_1\,(x_2) & \,_0^C D_x^\beta \phi_1\,(x_2) & \cdots & \,_0^C D_x^\beta \phi_1\,(x_2) \\ \,_0^C D_x^\beta \phi_2\,(x_3) & \,_0^C D_x^\beta \phi_2\,(x_3) & \cdots & \,_0^C D_x^\beta \phi_2\,(x_3) \\ \vdots & \vdots & \cdots & \vdots \\ \,_0^C D_x^\beta \phi_N\,(x_{N-1}) & \,_0^C D_x^\beta \psi_N\,(x_{N-1}) & \cdots & \,_0^C D_x^\beta \phi_N\,(x_{N-1}) \end{bmatrix}$$

Similarly, applying Eq. (4) on Eq. (2) we get

$$M_2 \, C \, L_2 = f(x),$$
$$M \, C \, L_3 = g_1(t), \qquad\qquad (9)$$
$$M \, C \, L_4 = g_2(t).$$

where

$$L_2 = \begin{bmatrix} \phi_1(x_1)\phi_2(x_1)\ldots\phi_N(x_1) \\ \phi_1(x_2)\phi_2(x_2)\ldots\phi_N(x_2) \\ \vdots \cdots \vdots \\ \phi_1(x_N)\phi_2(x_N)\ldots\phi_N(x_N) \end{bmatrix},$$

$L_3 = \left[\phi_1{}'(x_1)\phi_2{}'(x_1)\ldots\phi_N{}'(x_1) \right]^T, L_4 = \left[\phi_1{}'(x_N)\phi_2{}'(x_N)\ldots\phi_N{}'(x_N) \right]^T$ and $M_2 = \left[T_1(t_1) \, T_2(t_1) \ldots T_n(t_1) \right]$.

Collocating Eq. (9) will produce $(N + 2n - 2)$ equations. Combining these equations with Eq. (8) will result in Nn equations.

For the solution of Eq. (8), we convert it in a handy form using Kronecker product

$$\left\{ \rho_t c_t \left(L^t \otimes M_1 \right) - k \left(L_1{}^t \otimes M \right) + W_b c_b \left(L^t \otimes T \right) \right\} vec(c) = vec \left(W_b c_b T_a + q_{met} \right),$$

which can be formed as

$$A_1 C = F_1 \qquad\qquad (10)$$

where $(N - 2)(n - 1) \times Nn$ is the size of matrix A_1. $vec(c)$ is acquired by putting the columns of C on top of one another and is of size $Nn \times 1$ [1]. Size of vector F_1 is $(N - 2)(n - 1) \times 1$.

An expression for initial and boundary conditions can be given as

$$\left(L_2^t \otimes M2 \right)]c = T_a,$$
$$\left(L_3^t \otimes M \right) c = q_0$$
$$\left(L_4^t \otimes M \right) c = 0$$

that can be rewritten as

$$A_2 C = F_2,$$
$$A_3 C = F_3, \qquad\qquad (11)$$
$$A_4 C = F_4,$$

where matrix A_2 has the dimension $N \times Nn$. Size of vector F_2 is $N \times 1$. Matrices A_3 and A_4 have equal dimension i.e. $(n - 1) \times Nn$. Similarly, $(n - 1) \times 1$ is the size of vectors F_3 and F_4.

The resulting system involving Eqs. (10) and (11) is given by

$$AC = F \tag{12}$$

where $A = [A_1, A_2, A_3, A_4]^t$ is of dimension $Nn \times Nn$. $Nn \times 1$ is the size of the vector $F = [F_1, F_2, F_3, F_4]^t$.

The coefficient C can be obtained by solving the system given in (12). Use of this C in Eq. (4) will produce expected approximation for $T(x, t)$ which is the solution of the problem given in (1).

5 Results and Discussion

In this study, we have considered the following values of parameters [2] $L = 0.02\,\text{m}$, $T_a = 37\,°\text{C}$, $q_0 = 5000\,\text{W/m}^2$, $\rho_t = 1050\,\text{kg/m}^3$, $\rho_b = 1050\,\text{kg/m}^3$, $q_{met} = 368.1\,\text{W/m}^3$, $W_b = 0.5\,\text{kg/m}^3$, $c_b = 3770\,\text{JC/kg}$, $c_t = 4180\,\text{JC/kg}$ and $K = 0.5\,\text{WC/m}$

Figure 1 represents the temperature profile along the distance for different time. It demonstrates the comparison and verification of numerical solutions obtained by us and analytic solution by Shih et al. [10]. It also shows the numerical solution of FPBE obtained for $\alpha \to 1$, $\beta \to 2$.

Fig. 1 Comparison of numerical solution and analytic solution

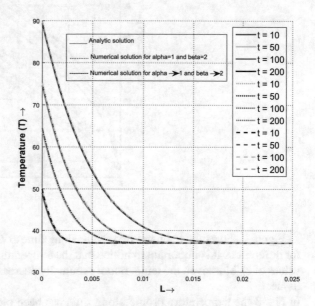

Fig. 2 Temperature
variation along time for
different α

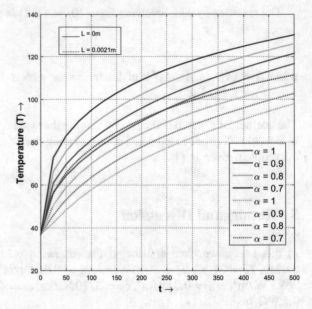

Fig. 3 Temperature
variation along time for
different β

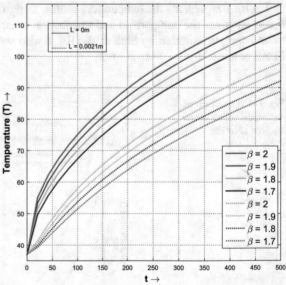

Figure 2 shows the temperature profile along time at $L = 0$ m and $L = 0.0021$ m
for different α. It is important to notice that the temperature is uplifted for decreased
values of α. Moreover, the temperature eventually decreases for increasing length of
tissue.

In Fig. 3 the temperature profile along time has been plotted for the different β at
$L = 0$ m and $L = 0.021$ m. Here the opposite behaviour of temperature is observed

Fig. 4 Temperature variation along time for different α and β

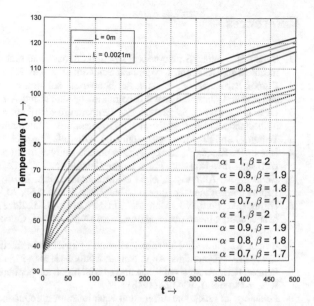

compared to Fig. 1 i.e. the temperature decreases for smaller values of β. Further, increasing depth of skin tissue is responsible for decrement in temperature.

Effect of simultaneous fractional values of α and β on temperature profile is analysed in Fig. 4. In this case also we find the same behaviour as Fig. 1 i.e. temperature gets upward elevation for decreasing fractional values of α and β. Further, the temperature decreases for increasing the length of tissue.

6 Conclusion

In the present study, the temperature profile of FPBE along time with constant heat flux has been examined. We observed anomalous diffusion for fractional values of α which is further compared with normal diffusion. Moreover, It is ascertained that the temperature keeps increasing for decreasing fractional values of α and the temperature goes down for smaller fractional values of β. Also for simultaneous fractional values of α and β the temperature increases and the temperature decreases if the depth of tissue increases.

Acknowledgements Conflict of Interest The first author is thankful to S. V. National Institute of Technology for providing financial support during the preparation of this manuscript.

References

1. Bansu, H., Kumar, S.: Numerical solution of space and time fractional telegraph equation: a meshless approach. Int. J. Nonlinear Sci. Numer. Simul. **20**, 325–337 (2019)
2. Damor, R., Kumar, S., Shukla, A.: Numerical solution of fractional bioheat equation with constant and sinusoidal heat flux coindition on skin tissue. Am. J. Math. Anal. **1**, 20–24 (2013)
3. Damor, R., Kumar, S., Shukla, A.: Temperature distribution in living tissue with fractional bioheat model in thermal therapy. In: Proceedings of International Conference on Advances in Tribology and Engineering Systems, pp. 493–498. Springer (2014)
4. Damor, R., Kumar, S., Shukla, A.: Parametric study of fractional bioheat equation in skin tissue with sinusoidal heat flux. Fract. Differ. Calculus **5**, 43–53 (2015)
5. Ezzat, M.A., AlSowayan, N.S., Al-Muhiameed, Z.I., Ezzat, S.M.: Fractional modelling of Pennes bioheat transfer equation. Heat Mass Transfer **50**, 907–914 (2014)
6. Ferrás, L.L., Ford, N.J., Morgado, M.L., Nóbrega, J.M., Rebelo, M.S.: Fractional Pennes bioheat equation: theoretical and numerical studies. Fract. Calculus Appl. Anal. **18**, 1080–1106 (2015)
7. Kumar, S., Piret, C.: Numerical solution of space-time fractional PDEs using RBF-QR and Chebyshev polynomials. Appl. Numer. Math. **143**, 300–315 (2019)
8. Pennes, H.H.: Analysis of tissue and arterial blood temperatures in the resting human forearm. J. App. Physiol. **1**, 93–122 (1948)
9. Podlubny, I.: Fractional differential equations: an introduction to fractional derivatives, fractional differential equations, to methods of their solution and some of their applications, vol. 198. Elsevier (1998)
10. Shih, T.C., Yuan, P., Lin, W.L., Kou, H.S.: Analytical analysis of the pennes bioheat transfer equation with sinusoidal heat flux condition on skin surface. Med. Eng. Phys. **29**, 946–953 (2007)

A Dynamic Finite Element Simulation of the Mitral Heart Valve Closure

Kamran Hassani

1 Introduction

The operation of the mitral valve (MV) is a crucial issue for the researchers. The valve works completely complex during the cardiac cycle there is interaction between the valve's structures [1]. The MV is located between the left atrium and left ventricle, and its duty is to regulates the flow between the atrium and the ventricle [2]. It includes four main components namely the leaflets, the papillary muscles, the chordae tendineae, and the annulus [3]. When MV is opened then it lets blood flow from the left atrium to left ventricle which is named diastole phase. During systole, the MV is close to prevent blood backflow into the atrium [4].

During closuring of the MV, a different set of pressures from the left atrium and left ventricles are being applied to the leaflets. The calculation of these stresses and strains experimentally is not achievable. Therefore, the application of numerical simulations, such as finite element method (FEM), are highly preferred. A finite element method model can connect the stresses and deformations to adaptions of leaflets mechanical response to organ-level. Various numerical models simulated the mechanical function of the MV, using computational fluid dynamics (CFD) [5, 6], FE [7, 8], and fluid–structure interaction (FSI) [9, 10]. So far, in spite of a wide range of clinical and biomechanical data in regard of the complex mechanical function of the normal MV, there is still a lack of information about the stresses and strains occurring during leaflets closuring [11, 12]. Therefore, we aimed at developing a simplified three-dimensional (3D) patient-specific FE model of the MV and papillary muscles on the MV closing behavior. The pressures from the left atrium and left ventricles were dynamically applied to the leaflets wall, and the resulted stresses and strains in the leaflets and papillary muscles were obtained.

K. Hassani (✉)
Department of Biomedical Engineering, Science and Research Branch, Islamic Azad University, Tehran, Iran
e-mail: k.hasani@srbiau.ac.ir

© Springer Nature Switzerland AG 2021
C. T. Lim et al. (eds.), *17th International Conference on Biomedical Engineering*, IFMBE Proceedings 79, https://doi.org/10.1007/978-3-030-62045-5_3

2 Models and Methods

A 3D FE model of the heart mitral valve and papillary muscles were reconstructed according to the patient-specific data [1, 13]. The structure of the model as well as the dimensions are displayed in Fig. 1. The leaflets, including the anterior, posteriors 1, 2, and 3, as well as commissural, were separately designed in Solidworks (Dassault Systèmes, Vélizy-Villacoublay, France) according to the data summarized in Table 1. The leaflets were then imported into Abaqus (Dassault Systèmes, Vélizy-Villacoublay, France) for assembly, materials, loading, and boundary conditions assignment. Extracting the exact geometry of the MV along with its chordal structure is very hard and the previous models have been simplified [14]. The same process has been done here for papillary muscles, as in our model only chordae tendons were modeled and were precisely connected to the leaflets according to the anatomical features of the heart mitral valve leaflets [1, 13]. The chordal structure is an essential section of the MV and could not be simplified. A cardiac cycle takes ~800 ms [15], and closuring phase of that lasts ~200 ms. Since this study was aimed at using FE model to dynamically calculate the stresses and strains in the MV leaflets in

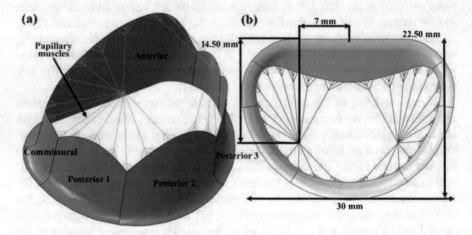

Fig. 1 The **a** FE model of the heart mitral valve, including the position of each leaflet and papillary muscles. The **b** dimensions of the model

Table 1 The dimensions of the FE model of the mitral valve leaflets

Leaflet	Annular length (mm)	Leaflet height (mm)	Leaflet area (mm^2)
Anterior	28.90	23.40	444.30
Commissural	7	8.70	46.10
Posterior 2	17.40	13.80	177.70
Posterior 1 and 3	12.70	11.20	111.40

closuring phase of the valves, only 200 ms of a cardiac cycle was simulated. During the closure phase of the valves, two pressures against each other are being applied on the leaflets, including one from the ventricles and the other one from the atrium. The distributions and profiles of the applied pressures from the ventricle and atrium during the closuring of the valve are indicated in Fig. 2. Due to a higher pressure from the ventricle (~15 kPa or 112.50 mmHg) compared to the atrium (~1.90 kPa or 14.25 mmHg), closuring phase of the valves was achievable.

The homogenous-isotropic linear mechanical properties here were employed to address the mechanical properties of the MV leaflets and papillary muscles under the applied loads. The mitral valve leaflets and papillary muscles were considered to be incompressible (~0.49) [15, 16] with the elastic modulus of 3 and 40 MPa, respectively [17]. The leaflets were simulated as shell element (S4R) with the thickness

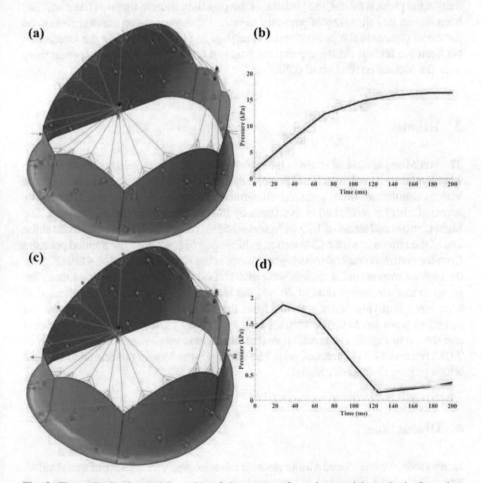

Fig. 2 The **a** distribution and **b** profile of the pressure from the ventricle to the leaflet valve. Similarly, The **c** distribution and **d** profile of the pressure from the atrium to the leaflet valve

and area of 0.75 mm and 0.6 mm^2 while papillary muscles were considered as beam element to diminish the simulation time. The models commonly take advantage of the chordal beam-type structure to justify the use of beam elements (i.e., have no bending degrees of freedom) and the thinness of the leaflets to justify the use of shell elements [18–21]. Mesh density analyses have also been performed in Abaqus to find the most suitable number of elements to not only diminish the simulation time but also have accurate numerical outcomes (data not reported here). The number of elements for the whole model was 11,865. Boundary condition has a key asset in a numerical simulation as it has to mimic the realistic anatomical behavior of the organ in a model [22–25]. Here, to fulfil this objective, the tip of the papillary muscles were fixed in all directions and angles. It was assumed that the papillary muscle contraction and displacement offset each other during the cardiac cycle so that, in the present model, the position of the papillary muscle tips was fixed as it has been shown that the effect of papillary muscle displacement and annular motion on the stress pattern at the peak pressure is negligible [17]. To simulate the interaction between the leaflets during the closure phase, a contact was defined between them with the friction coefficient of 0.50.

3   Results

The von Mises stress and strain in the leaflets of the MV through the simulation time were dynamically calculated and presented in Fig. 3. The structure of the valve under various simulations times were also illustrated in the inset of each panel. The results revealed a higher stress and higher stress by the passage of the simulation time. The highest stress and strain of 1.79 MPa and 49.51%, respectively, were observed at the end of the simulation time (200 ms) as exhibited in Fig. 4. Since the applied pressure from the ventricle reached to its highest value at the simulation time of ~130 (Fig. 3), the highest stresses in the leaflets were also shifted to a higher simulation time. That is, up to the simulation time of 28 ms, the leaflets were still have not contacted to each other implying that a pressure from the ventricle still is working against the atrium to close the MV. The stresses in the papillary muscles were also computed and shown in Fig. 5. The results revealed the highest von Mises stress and strain of 7.09 MPa and 14.21%, respectively. The stresses were mostly concentrated in those which pulling the anterior leaflet.

4   Discussions

In this study, we introduced a finite element model to study the closure of mitral valve, which included a simplified 3D structure of a patient-specific model reconstructed from high-resolution CT images (Fig. 1) [1, 13]. Measuring the In *vivo* force is an essential point for validation and adjust of material parameters in the models. Some

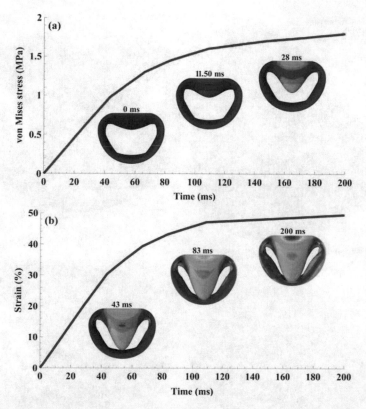

Fig. 3 The **a** von Mises stress versus simulation time and **b** strain versus simulation time in the mitral leaflet valves. The structure of the mitral valve under various simulation times are also presented in the inset of each panel

models are able to study the issues which could not be investigated experimentally, for example, in order to satisfy the static equilibrium,when the force is decreased in one side of the valve, it shall be increased in another [26]. Therefore, having a boundary condition which enables to mimic the realistic physiological response of the MV leaflets during closure is the first and foremost. Here, the pressures from the left ventricle and left atrium were applied on the leaflets wall as a function of time (Fig. 2), and thereafter, the resulted stresses and strains in the leaflets and papillary muscles were calculated.

The results revealed the highest stress in the anterior leaflet as 1.79 MPa at the end of the simulation time (Figs. 3 and 4). The stresses of 1.60 MPa [27], 1.51 MPa [28], and 1.50 MPa [29] in the leaflets wall. In addition, the deformed configuration of the closed valve in this study (Fig. 3) was in good agreement with Morganti et al., although they employed a nonlinear Mooney-Rivlin hyperelastic material model to address the mechanical properties of the MV leaflets [27].

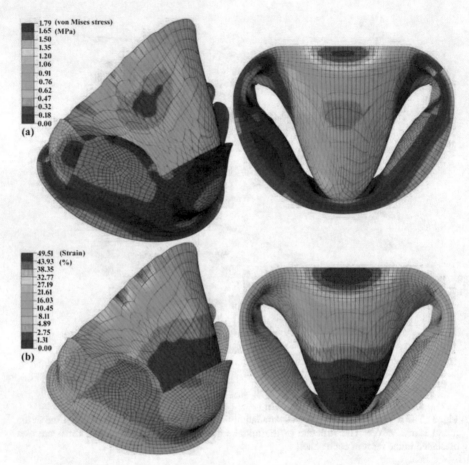

Fig. 4 The **a** von Mises stress and **b** strain contours in the mitral valve leaflets from the isometric and bottom views at the end of the smulation time (200 ms)

In addition, the strains of 47.10% [30] and 47% [31] were observed in the leaflet walls, which are in good agreement with that of the present study of 49.51% (Fig. 4). The highest strains in our results were observed in the apex of the anterior leaflets as also reported by [32, 33]. The calculated strains in the papillary muscles were found to be 15% [28] and 12.50% [30]. Here, the stress and strain of 7.09 MPa and 14.21%, respectively, were observed in the papillary muscle (Fig. 5).

The results reported herein require careful interpretation. In particular, several factors must be considered when applying these findings to clinical practice. In the model, a linear elastic behavior of the leaflets and papillary muscles were assumed. Biaxial strip tests show that both leaflets have a non-linear, elastic stress–strain behavior [34]. Despite linear elasticity being currently used for the simulation of leaflet valves [35, 36], due to computational cost saving and convergence requirements, further studies will need to include this feature.

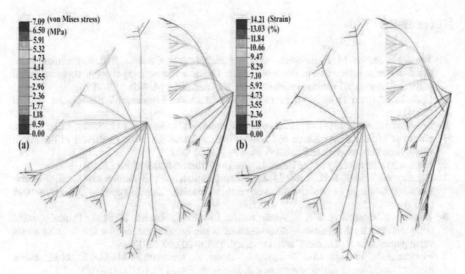

Fig. 5 The **a** von Mises stress and **b** strain contours in the papillary muscles of the mitral valve at the end of the smulation time (200 ms)

5 Conclusion

This study was aimed at conducting a 3D FE study to analyze the stresses and strain in the closuring phase of the heart MV. To do that, the patient-specific structure of the heart valve leaflets and papillary muscles were established and subjected to the dynamic pressures applied from the left ventricle and left atrium during a cardiac cycle. The results revealed higher stresses and strains in the anterior leaflet compared to the other ones. The papillary muscles connecting to the anterior leaflet were also showed higher stresses and strains compared to those connected to other leaflets.

The results have implications for providing a comprehensive information for the medical and biomechanical experts in regard of the stresses and deformations in the mitral valve leaflets and papillary muscles during the closuring of the leaflets of the heart valve.

Conflicts of Interest
None declared.

Funding This research received no specific grant from any funding agency in the public, commercial, or not-for-profit sectors.

References

1. Toma, M., Jensen, M.Ø., Einstein, D.R., Yoganathan, A.P., Cochran, R.P., Kunzelman, K.S.: Fluid–structure interaction analysis of papillary muscle forces using a comprehensive mitral valve model with 3D chordal structure. Ann. Biomed. Eng. **44**, 942–953 (2016)
2. Opie, L.H.: Heart Physiology: From Cell to Circulation. Lippincott Williams and Wilkins (2004)
3. Lee, C.H., Rabbah, J.P., Yoganathan, A.P., Gorman, R.C., Gorman, J.H., Sacks, M.S.: On the effects of leaflet microstructure and constitutive model on the closing behavior of the mitral valve. Biomech. Model Mechanobiol. **14**, 1281–1302 (2015)
4. Katz, A.M.: Physiology of the Heart. Lippincott Williams and Wilkins (2010)
5. Randles, A., Frakes, D.H., Leopold, J.A.: Computational fluid dynamics and additive manufacturing to diagnose and treat cardiovascular disease. Trends Biotech. **35**, 1049–1061 (2017)
6. Youssefi, P., Gomez, A., He, T., Anderson, L., Bunce, N., Sharma, R., et al.: Patient-specific computational fluid dynamics—assessment of aortic hemodynamics in a spectrum of aortic valve pathologies. J. Thorac. Cardiovas. Surg. **153**(8–20), e3 (2017)
7. Morgan, A.E., Pantoja, J.L., Weinsaft, J., Grossi, E., Guccione, J.M., Ge, L., et al.: Finite element modeling of mitral valve repair. J. Biomech. Eng. **138**, 021009 (2016)
8. Choi, A., Rim, Y., Mun, J.S., Kim, H.: A novel finite element-based patient-specific mitral valve repair: virtual ring annuloplasty. Bio-Med. Mater. Eng. **24**, 341–347 (2014)
9. Kamensky, D., Hsu, M.C., Schillinger, D., Evans, J.A., Aggarwal, A., Bazilevs, Y., et al.: An immersogeometric variational framework for fluid–structure interaction: application to bioprosthetic heart valves. Comput. Methods Appl. Mech. Eng. **284**, 1005–1053 (2015)
10. Govindarajan, V., Kim, H., Chandran, K., McPherson, D.: Fluid structure interaction simulation to improve evaluation of mitral valve repair. J. Am. Coll. Cardiol. **71**, A1977 (2018)
11. Baillargeon, B., Rebelo, N., Fox, D.D., Taylor, R.L., Kuhl, E.: The living heart project: a robust and integrative simulator for human heart function. Eur. J. Mech. Solids **48**, 38–47 (2014)
12. Dallard, J., Labrosse, M.R., Sohmer, B., Beller, C.J., Boodhwani, M.: Investigation of raphe function in the bicuspid aortic valve and its influence on clinical criteria–a patient- specific finite-element study. Int. J. Numer. Method Biomed. Eng. e3117 (2018)
13. Baillargeon, B., Costa, I., Leach, J.R., Lee, L.C., Genet, M., Toutain, A., et al.: Human cardiac function simulator for the optimal design of a novel annuloplasty ring with a sub-valvular element for correction of ischemic mitral regurgitation. Cardiovasc. Eng. Tech. **6**, 105–116 (2015)
14. Jensen, M.Ø., Fontaine, A.A., Yoganathan, A.P.: Improved in vitro quantification of the force exerted by the papillary muscle on the left ventricular wall: three-dimensional force vector measurement system. Ann. Biomed. Eng. **29**, 406–413 (2001)
15. Karimi, A., Razaghi, R.: Interaction of the blood components and plaque in a stenotic coronary artery. Artery. Res. **12**, 47–61 (2018)
16. Karimi, A., Sera, T., Kudo, S., Navidbakhsh, M.: Experimental verification of the healthy and atherosclerotic coronary arteries incompressibility via digital image correlation. Artery Res. **16**, 1–7 (2016)
17. Votta, E., Maisano, F., Soncini, M., Redaelli, A., Montevecchi, F.M., Alfieri, O.: 3-D computational analysis of the stress distribution on the leaflets after edge-to-edge repair of mitral regurgitation. J. Heart. Valve Dis. **11**, 810–822 (2002)
18. Einstein, D.R., Del Pin, F., Jiao, X., Kuprat, A.P., Carson, J.P., Kunzelman, K.S., et al.: Fluid–structure interactions of the mitral valve and left heart: comprehensive strategies, past, present and future. Int. J. Numer. Method Biomed. Eng. **26**, 348–380 (2010)
19. Lau, K., Diaz, V., Scambler, P., Burriesci, G.: Mitral valve dynamics in structural and fluid–structure interaction models. Med. Eng. Phys. **32**, 1057–1064 (2010)
20. Mansi, T., Voigt, I., Georgescu, B., Zheng, X., Mengue, E.A., Hackl, M., et al.: An integrated framework for finite-element modeling of mitral valve biomechanics from medical images: application to MitralClip intervention planning. Med. Image Anal. **16**, 1330–1346 (2012)

21. Sacks, M.S.: Incorporation of experimentally-derived fiber orientation into a structural constitutive model for planar collagenous tissues. J. Biomech. Eng. **125**, 280–287 (2003)
22. Karimi, A., Navidbakhsh, M., Razaghi, R.: Plaque and arterial vulnerability investigation in a three-layer atherosclerotic human coronary artery using computational fluid-structure interaction method. J. Appl. Phys. **116**, 064701–064710 (2014)
23. Karimi, A., Navidbakhsh, M., Razaghi, R., Haghpanahi, M.: A computational fluid-structure interaction model for plaque vulnerability assessment in atherosclerotic human coronary arteries. J. Appl. Phys. **115**, 144702–144710 (2014)
24. Karimi, A., Navidbakhsh, M., Yamada, H., Razaghi, R.: A nonlinear finite element simulation of balloon expandable stent for assessment of plaque vulnerability inside a stenotic artery. Med. Biol. Eng. Comput. **52**, 589–599 (2014)
25. Karimi, A., Shojaei, A., Razaghi, R.: Viscoelastic mechanical measurement of the healthy and atherosclerotic human coronary arteries using DIC technique. Artery Res. **18**, 14–21 (2017)
26. Rahmani, A., Rasmussen, A.Q., Honge, J.L., Ostli, B., Levine, R.A., Hagège, A., et al.: Mitral valve mechanics following posterior leaflet patch augmentation. J. Heart Valve Dis. **22**, 28 (2013)
27. Morganti, S., Auricchio, F., Benson, D.J., Gambarin, F.I., Hartmann, S., Hughes, T.J.R., et al.: Patient- specific isogeometric structural analysis of aortic valve closure. Comput. Methods Appl. Mech. Eng. **284**, 508–520 (2015)
28. Pham, T., Kong, F., Martin, C., Wang, Q., Primiano, C., McKay, R., et al.: Finite element analysis of patient-specific mitral valve with mitral regurgitation. Cardiovasc. Eng. Technol. **8**, 3–16 (2017)
29. Ge, L., Morrel, W.G., Ward, A., Mishra, R., Zhang, Z., Guccione, J.M., et al.: Measurement of mitral leaflet and annular geometry and stress after repair of posterior leaflet prolapse: virtual repair using a patient-specific finite element simulation. Ann. Thorac. Surg. **97**, 1496–1503 (2014)
30. Bloodworth, C.H., Pierce, E.L., Easley, T.F., Drach, A., Khalighi, A.H., Toma, M., et al.: Ex vivo methods for informing computational models of the mitral valve. Ann. Biomed. Eng. **45**, 496–507 (2017)
31. Sacks, M.S., He, Z., Baijens, L., Wanant, S., Shah, P., Sugimoto, H., et al.: Surface strains in the anterior leaflet of the functioning mitral valve. Ann. Biomed. Eng. **30**, 1281–1290 (2002)
32. Hsu, M.C., Kamensky, D., Xu, F., Kiendl, J., Wang, C., Wu, M.C.H., et al.: Dynamic and fluid–structure interaction simulations of bioprosthetic heart valves using parametric design with T-splines and Fung-type material models. Comput. Mech. **55**, 1211–1225 (2015)
33. Morgan, A.E., Pantoja, J.L., Grossi, E.A., Ge, L., Weinsaft, J.W., Ratcliffe, M.B.: Neochord placement versus triangular resection in mitral valve repair: a finite element model. J. Surg. Res. **206**, 98–105 (2016)
34. May-Newman, K., Yin, F.: A constitutive law for mitral valve tissue. J. Biomech. Eng. **120**, 38–47 (1998)
35. Kunzelman, K., Reimink, M., Cochran, R.: Flexible versus rigid ring annuloplasty for mitral valve annular dilatation: a finite element model. J. Heart Valve Dis. **7**, 108–116 (1998)
36. Cochran, R.P., Kunzelman, K.S.: Effect of papillary muscle position on mitral valve function: relationship to homografts. Ann. Thorac. Surg. **66**, S155–S161 (1998)

Choroid Segmentation in Optical Coherence Tomography Images Using Deep Learning

Ruchir Srivastava, Ee Ping Ong, and Beng-Hai Lee

1 Introduction

Analysis of the choroid is important to assess diseases such as glaucoma, age-related macular degeneration (AMD), serous chorioretinopathy and choroidal melanoma, which accompany choroidal changes [5]. A common imaging modality to visualize the retinal layers and the choroid is optical coherence tomography (OCT). An OCT scan comprises of hundreds of slices (called B-scans) which makes manual analysis of a scan highly labour intensive. This is especially of a concern when most of the examined scans turn out to be normal. Moreover, manual examination can be highly subjective. In order to save the labour and avoid the subjectivity involved in manual analysis, automatic analysis of the choroid can be highly beneficial for clinical applications.

The first step in automatic analysis of the choroid is to segment it in OCT B-scans (referred to henceforth as 'images'). Even though there has been extensive research in automatic retinal layer segmentation, only a few works segment the choroid. Existing choroid segmentation methods include graph search [13], statistical models [9] and other image processing methods [2, 5]. Another possible method is to use deep learning which has shown better performance than conventional machine learning techniques for many image segmentation tasks. Deep networks can learn task-specific high-level discriminative features. Sui et al. [12] proposed to use a convolutional neural network (a deep learning tool) to provide a probability map which was used by graph search to segment the choroid. A similar approach was used by Alonso-Canero et al. [3] who also performed an initial segmentation of the choroid using CNN and refined it using graph search. Hassan et al. [8] segmented the retinal layers and the choroid using a tensor-based method and refined it using CNN.

R. Srivastava (✉) · E. P. Ong · B.-H. Lee
Institute for Infocomm Research, Singapore, Singapore
e-mail: srivastavar@i2r.a-star.edu.sg
URL: https://www.a-star.edu.sg/i2r

© Springer Nature Switzerland AG 2021
C. T. Lim et al. (eds.), *17th International Conference on Biomedical Engineering*,
IFMBE Proceedings 79, https://doi.org/10.1007/978-3-030-62045-5_4

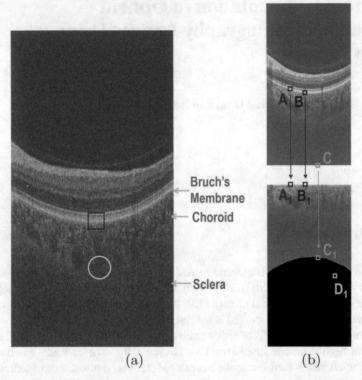

(a)   (b)

Fig. 1 (Best viewed in colour) **a** An OCT image showing the choroid which is bound by the Bruch's membrane from the top and sclera from the bottom. The upper boundary within the square is clear while the lower boundary within the circle is vague. This makes choroid segmentation using boundary-based approaches difficult. We propose to segment the choroid region as a whole instead of detecting its boundaries. **b** An illustration of the procedure for removing the retinal layers. Points A, B, and C in the original image are mapped to A1, B1, and C1, respectively in the new image. Since there is no correspondence for the point D1 in the new image, its pixel value is put to zero (black)

All of these works using deep learning performed a boundary-based segmentation of the choroid. Such an approach tries to segment the boundaries enclosing the choroid using the edge information present at these boundaries. This approach would work well in cases where the boundaries are well defined. For the choroid, the boundary between the retina and the choroid (called the Bruch's membrane or BM in short) is well defined. However, the boundary between the choroid and sclera or the choroido-scleral junction (CSJ) is vague in many cases (Fig. 1a). This was also experimentally observed in [3] where detecting CSJ was less accurate as compared to detecting retinal boundaries. In such cases with a vague CSJ, using a region-based segmentation approach can be more accurate as it can characterize the unique texture of the choroid (Fig. 1a). A region-based approach learns the characteristics of the region to be segmented instead of just the information near the boundaries. Such an approach has been used by Venhuizen et al. [14] for segmentation of the retina using a deep architecture called U-Net [10]. U-Net uses a region-based approach called semantic segmentation in computer vision. Semantic segmentation involves classifying each pixel in an image into one of the many object classes like road, sky and tree in an object detection task.

In this work we propose to extend the work in [14] to segment the choroid. However, in segmenting the choroid there are two approaches possible. First approach can be to remove the retina from the images before performing choroid segmentation. This can reduce the confusion caused by the retinal layers in the choroid segmentation task. In this approach, the BM is detected and based on this detection, the layers above the BM are cropped out. There is a limitation of this approach since the accuracy of choroid segmentation will in turn depend on the accuracy of BM detection. However, usually the BM is a well-defined layer so the loss of accuracy due to BM detection can be compensated by a higher accuracy for choroid segmentation. The second approach to choroid segmentation does not remove the retinal layers and directly perform the choroid segmentation as a semantic segmentation task. In this work, we compare both the above mentioned approaches for the choroid segmentation problem.

Contributions: The contributions of the proposed work are two-fold. Firstly, we use a region-based approach for choroid segmentation. This captures the unique texture of the choroid which differentiates it from other retinal layers and the sclera. Secondly, we experimentally examine the effect of removing retinal layers before choroid segmentation. This is important to obtain accurate choroid segmentation. Moreover, the findings can give insight to other segmentation tasks involving multiple layers (such as retinal layers or skin layers).

2 Method

A flowchart of the proposed method is shown in Fig. 2. The proposed method is a supervised method with training and testing stages and takes OCT images as input. Each image is preprocessed to normalize the intensity across the different images after which the BM is detected so as to remove the retinal layers. Thereafter, the images from the training set (or training images) are divided into smaller patches. Data augmentation is performed to increase the amount of training data. Resulting patches are used to train U-Net.

From the test images, patches are extracted and fed along with their ground truth labels to the trained U-Net for choroid segmentation. Test images are then processed similar to the training images and patches are extracted from them. These patches are fed to the trained U-Net which then outputs a binary segmentation map with choroid pixels as 1 and others 0. The resulting segmentation map is post-processed to obtain the final segmentation result. We call this method CS-BM (Choroid Segmentation after flattening BM). For the second method of not removing the retinal layers (called

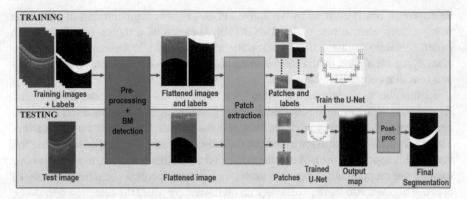

Fig. 2 A flowchart of the proposed method for choroid segmentation (CS-BM)

CS), all the other steps are the same as CS-BM but it does not involve BM detection
and removal of the retinal layers. Hence to avoid any confusion, we present the details
of only CS-BM in the next section.

2.1 Preprocessing and BM Detection

Input images from both training and test datasets were intensity normalized using the
method in [7]. The image was rescaled such that the intensity values were between
0 and 1. If I_{max} and I_{min} denote the maximum and minimum intensity values in the
image, the rescaled intensity, I_r is given as:

$$I_r = \frac{I_0 - I_{min}}{I_{max} - I_{min}} \tag{1}$$

where I_0 is the intensity before rescaling. A median filter, 20 pixels high and 2 pixels
wide was applied on the resulting image for noise removal. Thereafter, any intensity
larger than $1.05 \times I_m$ was set to $1.05 \times I_m$ where I_m denotes the maximum intensity
in the rescaled and noise removed image.

After preprocessing, for the test images, BM was detected using an unsupervised
graph-based method originally used for skin surface segmentation [11] (note that
for the training images, BM detection was not needed as we used the ground truth,
instead). We modified the cost function of this method such that the computed cost
was lowest at the BM. This cost function utilized two observations, (1) there is a
strong edge at the BM, and (2) there is a bright to dark transition at the BM when
going from retina to the choroid. Once the cost was defined, BM was detected as the
surface with the lowest cost using the max-flow/min-cut algorithm [4]. BM detection
was followed by cropping out the retinal region from the images as explained in the
following section.

2.2 Choroid Segmentation

Choroid segmentation was performed using U-Net which is a deep neural network found to be successful in biomedical image segmentation tasks. The structure of U-NET is shown in Fig. 3.

In this work, U-Net was designed to accept patches of size 256×256 pixels along with a binary label image with only the choroid pixels as white. The original images were of size 512×992 pixels. Before extracting these patches from the training images and the labels, we used the ground truth for BM location to remove the retinal layers. In order to remove the retinal layers, portion of the image above the BM was cropped out. This resulted in a new image where the BM appears flat at the top. Figure 1b illustrates this process where the original image is shown at the top while the new image with retinal layers removed is shown below it. The points A, B, and C in the original image are mapped to A1, B1, and C1 in the new image. However, the point D1 does not have any correspondence in the original image (since we enforce the new image to be of the same size as the original image). For all such points, the image intensity was put to zero (black).

Thereafter, the images were divided into 28 square patches along with their labels. Each patch was of size 128×128 pixels. This left out some portion of the image in the bottom which was acceptable since we observed that this region did not con-

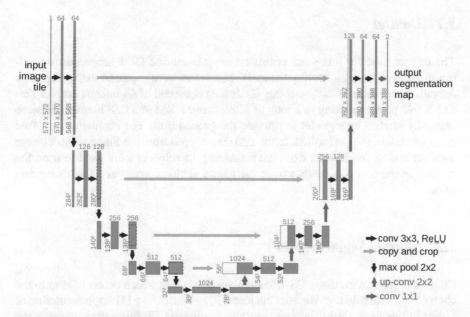

Fig. 3 The structure of U-Net for an input image size of 572×572 pixels used in the original work [10]. For this work we use input images of size 256×256 so the filter output sizes will change accordingly throughout the network

tain choroid. Out of these patches, we removed patches containing less than 20% or more than 80% of the choroid. This ensured a balance of choroid and non-choroid in the patches. These patches were resized to 256×256 pixels to match the input requirements for U-Net. To further increase the training data size, data augmentation was performed. Each patch was rotated by $90°$. Both the original and rotated versions were flipped horizontally increasing the number of patches to four-fold. These patches were used to train the network.

To extract patches from the test images, the process was the same as for the training images just that BM was automatically detected (Sect. 2.1), data augmentation was not needed and only the patches in the bottom were excluded which were not square. The extracted patches were fed to the trained U-Net which outputted a score for each pixel indicating the probability that the pixel belongs to the choroid. Results of the patches were collated to get the prediction for the full image. Thereafter, the probability values were thresholded to yield a binary image containing many separate blobs or connected components. In order to remove false detections, only the largest connected component was kept and the rest were ignored. Retinal layers were added to the resulting binary image to get the final segmentation.

3 Results and Discussion

3.1 Dataset

The dataset used for this work consisted macula-centred OCT scans corresponding to 20 eyes from 20 healthy subjects. The scans were captured using Topcon's Atlantis swept source OCT machine. Each scan consisted of 64 images each of size 512×992 pixels resulting in a total of 1280 images. BM and CSJ boundaries were manually marked by a grader to provide the ground truth. For evaluation, a 4 fold cross validation was performed. In the first run, images from the first 5 subjects were used for testing and images from the remaining 15 subjects were used to train the system. In the next run, images from the next 5 subjects were used for testing and so on.

3.2 Choroid Segmentation

Choroid segmentation using CS-BM involves predicting which pixels belong to the choroid and which do not. We used the Keras [6] + Tensorflow [1] implementation of U-Net in this work (https://github.com/zhixuhao/unet). Training parameters were: 50 epochs,batch size 16, learning rate 0.0001 with Adam optimizer and binary cross-entropy loss. On a Linux 64-bit PC with 125 GB RAM and Titan X (Pascal) GPU, the training for each run took around 9 h. Testing was performed as mentioned in

Table 1 Results of choroid segmentation using the proposed method (CS-BM) compared to not removing retinal layers before segmentation (CS) and edge-based segmentation method using graph search (GS)

	CS-BM	CS	GS
Run 1	0.82 (0.10)	0.79 (0.09)	0.56 (0.17)
Run 2	0.90 (0.06)	0.85 (0.07)	0.48 (0.23)
Run 3	0.84 (0.10)	0.78 (0.09)	0.52 (0.25)
Run 4	0.85 (0.08)	0.84 (0.07)	0.48 (0.22)
Mean	0.85	0.81	0.51

Values shown are mean values of IoU for each run with standard deviation in braces

Sect. 2.2. A threshold of 0.5 was used to obtain the binary segmentation from the probability estimates. To evaluate the results of choroid segmentation, a commonly used metric for semantic segmentation, intersection over union (IoU) was used. IoU is defined as:

$$IoU = \frac{GT \cap Pred}{GT \cup Pred} \qquad (2)$$

where GT denotes the ground truth region for the choroid while $Pred$ denotes the predicted choroid region. Results of choroid segmentation are shown in Table 1 where we compare CS and CS-BM. In addition, we have also compared our work with an edge-detection based approach using graph search. The graph search method is the same method used for detecting BM (Sect. 2.1). However, we defined a new cost function so as to detect the CSJ. This cost function mainly uses the vertical image gradient. In addition, the retinal layers were removed as in CS-BM before detecting the CSJ. We denote the graph search method as GS in Table 1.

Compared to GS (IoU = 0.51), both CS-BM and CS perform much better. Upon further examination, we realized that this was mainly because of poor visibility of CSJ in many images. In such cases, the proposed texture-based approaches performed much better. Comparing the two methods using a texture-based approach, the values of IoU are higher for CS-BM (0.85) as compared to CS (0.81). Results are as expected and removal of retinal layers improves the choroid segmentation accuracy. We further investigated the reason for inferior performance of CS and found that although there were false positives above the BM for CS, these detections were removed when we only considered the largest connected component. The actual cause of performance degradation was false detections in the sclera region as shown in Fig. 4 third row. This is possible due to a more difficult classification task for CS where it had to distinguish the choroid from sclera and other retinal layers. Consequently, the learned model was not as accurate as CS-BM which effectively had to distinguish choroid from just the sclera. This finding may be extended to other segmentation tasks where a similar simplification of the classification task may lead to a better segmentation. However, this improvement in accuracy was reduced by the error in BM detection. We evaluated the BM detection accuracy using unsigned error which is the mean absolute error

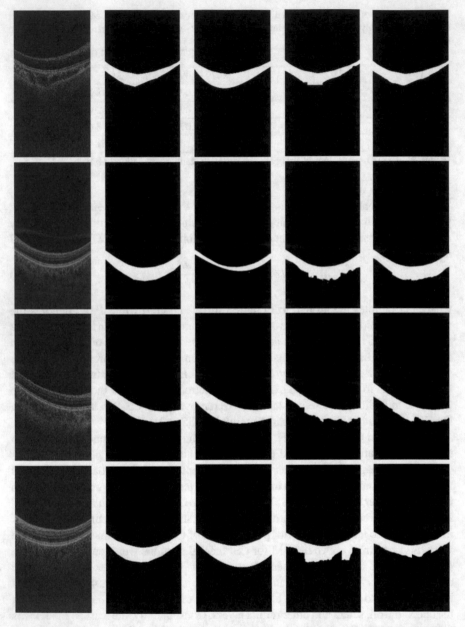

Fig. 4 Four sample results of choroid segmentation. From left to right are the original images, ground truths, results of GS, CS and the proposed method (CS-BM) respectively. The IoU values for GS are 0.733, 0.375, 0.748, and 0.642 for the four samples respectively from top to bottom while the corresponding values for CS are 0.913, 0.838, 0.885, and 0.794 and for CS-BM are 0.964, 0.903, 0.916, and 0.767

between predicted and actual location of the BM [3]. This error was found to be 2.2 pixels which, if reduced, can further improve the choroid segmentation accuracy.

Figure 4 shows examples of choroid segmentation where CS-BM is compared to CS along with GS and the ground truth. As observed in Fig. 4 first three rows, CS-BM segments the choroid more accurately than CS. However, in Fig. 4 last row, both CS and CS-BM are not as accurate and CS performs slightly better (IoU = 0.794) as compared to CS-BM (IoU = 0.767) since the CSJ was not clearly visible. In all the cases, GS achieves a smooth segmentation however, it is not as accurate as CS-BM. In the second row, GS fails to segment the CSJ and falsely detects another edge as the CSJ. This shows an example of a case where edge detection may fail due to strong edges apart from the CSJ. Even in such a case, both the region-based approaches perform significantly better with IoUs 0.838 and 0.903 for CS and CS-BM, respectively.

4 Conclusion

This paper presented a method to segment the choroid in OCT images using a region-based approach in which the choroid region as a whole is segmented. This utilizes the texture of the choroid. This is different from other approaches which only attempt to detect the boundary of the choroid using edge-detection methodologies. The proposed method uses U-Net, a deep learning network, for the segmentation task. We evaluated the proposed method on a dataset of 1280 images and compared it to an edge-based choroid segmentation method which uses graph search. The proposed method gave an intersection over union (IoU) of 0.85 while the graph search method gave an IoU of 0.51. Results indicate that utilizing the texture in the choroid can be useful especially in cases where choroidal boundaries may not be clear. This is especially useful in diseased cases where the boundaries are irregular and utilizing the choroidal texture may prove to be more accurate. In addition, we also assessed the effect of removing the retinal layers from the images before choroid segmentation. When retinal layers were not removed the obtained IoU was 0.81 which was slightly lower than the proposed method. The improvement in accuracy shows that it will help to remove retinal layers before choroid segmentation in OCT images. Note that the values of IoU are still not ideal for practical applications. This is because the results presented here are preliminary results as we first wanted to compare the two methods, CS and CS-BM. Future work involves attempts to improve this accuracy. With an improved accuracy, the proposed method can be use for further analysis of the choroid in diseases such as AMD and glaucoma. In such diseases, the boundaries are irregular and a region-based approach may prove to be more accurate.

References

1. Abadi, M., et al.: TensorFlow: large-scale machine learning on heterogeneous systems (2015). https://www.tensorflow.org/, software available from tensorflow.org
2. Alonso-Caneiro, D., Read, S.A., Collins, M.J.: Automatic segmentation of choroidal thickness in optical coherence tomography. Biomed. Opt. Exp. **4**(12), 2795–2812 (2013)
3. Alonso-Caneiro, D., Read, S.A., Hamwood, J., Vincent, S.J., Collins, M.J.: Use of convolutional neural networks for the automatic segmentation of total retinal and choroidal thickness in OCT images (2018)
4. Boykov, Y., Kolmogorov, V.: An experimental comparison of min-cut/max-flow algorithms for energy minimization in vision. IEEE Trans. Pattern Anal. Mach. Intell. **26**(9), 1124–1137 (2004)
5. Chen, Q., Fan, W., Niu, S., Shi, J., Shen, H., Yuan, S.: Automated choroid segmentation based on gradual intensity distance in HD-OCT images. Opt. Exp. **23**(7), 8974–8994 (2015)
6. Chollet, F., et al.: Keras. https://keras.io (2015)
7. Fang, L., Cunefare, D., Wang, C., Guymer, R.H., Li, S., Farsiu, S.: Automatic segmentation of nine retinal layer boundaries in oct images of non-exudative AMD patients using deep learning and graph search. Biomed. Opt. Exp. **8**(5), 2732–2744 (2017)
8. Hassan, T., Usman, A., Akram, M.U., Masood, M.F., Yasin, U.: Deep learning based automated extraction of intra-retinal layers for analyzing retinal abnormalities. In: 2018 IEEE 20th International Conference on e-Health Networking, Applications and Services (Healthcom), pp. 1–5. IEEE (2018)
9. Kajić, V., Esmaeelpour, M., Považay, B., Marshall, D., Rosin, P.L., Drexler, W.: Automated choroidal segmentation of 1060 nm OCT in healthy and pathologic eyes using a statistical model. Biomed. Opt. Exp. **3**(1), 86–103 (2012)
10. Ronneberger, O., Fischer, P., Brox, T.: U-Net: convolutional networks for biomedical image segmentation. In: International Conference on Medical Image Computing and Computer-Assisted Intervention. pp. 234–241. Springer (2015)
11. Srivastava, R., Yow, A.P., Cheng, J., Wong, D.W., Tey, H.L.: Three-dimensional graph-based skin layer segmentation in optical coherence tomography images for roughness estimation. Biomed. Opt. Exp. **9**(8), 3590–3606 (2018)
12. Sui, X., Zheng, Y., Wei, B., Bi, H., Wu, J., Pan, X., Yin, Y., Zhang, S.: Choroid segmentation from optical coherence tomography with graph-edge weights learned from deep convolutional neural networks. Neurocomputing **237**, 332–341 (2017)
13. Tian, J., Marziliano, P., Baskaran, M., Tun, T.A., Aung, T.: Automatic segmentation of the choroid in enhanced depth imaging optical coherence tomography images. Biomed. Opt. Exp. **4**(3), 397–411 (2013)
14. Venhuizen, F.G., van Ginneken, B., Liefers, B., van Grinsven, M.J., Fauser, S., Hoyng, C., Theelen, T., Sánchez, C.I.: Robust total retina thickness segmentation in optical coherence tomography images using convolutional neural networks. Biomed. Opt. Exp. **8**(7), 3292–3316 (2017). http://www.osapublishing.org/boe/abstract.cfm?URI=boe-8-7-3292

Design Concept for an Automated Lower Extremity Dressing Aid for Monoplegic and Elderly People

Allain Jessel Macas, Aaron Raymond See, Vu Trung Hieu, Yu-Yang Hsu, and Zheng-Kai Wang

1 Introduction

According to the World Stroke Organization, 15 million people worldwide suffer a stroke each year and 5.8 million people die from it. It has become one of the leading causes of long-term disability. About 60% of total stroke patients get permanently neural disabilities, which may impair normal behavioral functions [1].

One of the effects which is shown within the few months in patients after stroke is monoplegia, or the paralysis of one limb, usually an arm. The number of monoplegia patients is increasing rapidly with modernization, various kinds of accidents like industrial accidents and hereditary diseases like infantile paralysis [2]. Monoplegia occurs in 90% with the arm involved twice as often as the leg [3].

In this case, these patients would have a difficulty to take part in life's activities at home, school, work, and in the community. They are faced with the challenge to perform trivial tasks and daily routines on their own. These tasks include getting up from bed, taking a shower, dressing and undressing, and many others.

To eliminate these barriers and further increase the sense of independence of these individuals, assistive devices have been developed to address their needs especially on putting on and taking off clothes. Over the years, a variety of assistive device have been invented to aid individuals put on the upper and lower garments by themselves. One study about the adaptive device used by older adults with mixed disabilities showed that among all, the dressing devices were the most frequently issued piece of equipment [4]. In addition, another study showed that lower extremities dressing is one of the activities that confirms a higher percentage of individual with disabilities in

A. J. Macas · A. R. See (✉) · V. T. Hieu · Y.-Y. Hsu · Z.-K. Wang
Southern Taiwan University of Science and Technology, Tainan 71005, Taiwan
e-mail: aaronsee@stust.edu.tw

A. J. Macas
University of Science and Technology of Southern Philippines, 9000 Cagayan de Oro, The Philippines

© Springer Nature Switzerland AG 2021
C. T. Lim et al. (eds.), *17th International Conference on Biomedical Engineering*,
IFMBE Proceedings 79, https://doi.org/10.1007/978-3-030-62045-5_5

which they were unable to perform and for which help from others was required [5]. The same study also displayed a much higher percentage of needs of these specific devices than its actual usage. This means that the existing devices do not meet their needs. Another study explored the use of assistive devices for dressing by older persons with impairments and it showed that the group with both upper and lower extremity dressing difficulty reported the highest level of pain and scored the lowest on all measures of functional status and mental status. The most commonly used dressing devices were associated with lower extremity dressing. It also summarized the top reasons for dissatisfaction with dressing devices to be the perception that they don't need or use them, devices don't work well, and too difficult to use [6].

Many of the existing lower garment dressing aid today may take too much time to use and even require the patient to ask help from others in using them. Taking off the lower garment using some of these devices may demand the patient to bend and exert more effort than necessary.

On that note, this project aims to simulate an automated dressing aid of lower garments for people with monoplegia and elder people. For further research, the project's specific objectives are as follows: to help the patient dress or undress the lower garment in a convenient and efficient way; to be portable and light-weight enough to be moved around; and to create easier control for the patient to use it independently.

2 Design Criteria

2.1 Background Research

A study entitled "Adaptive Use by Older Adults with Mixed Disabilities" examined home equipment use over a three-month period by 13 elderly patients discharged from a hospital rehabilitation unit. Dressing devices were the most frequently issued category of equipment followed by the bathroom devices for this study sample. In addition, of the 23 dressing aids, 11(47%) were reported in frequent or consistent use.

Therapist also cited that the major reasons for nonuse of equipment such as the dressing equipment are the lack of knowledge as to appropriate use, and what to do with the broken of defective devices. The research demonstrated that the ease of use of a device and how it fits into the personal lifestyle and care management plan of the individual directly effects frequency of use [4]. Some also identified that in using a device, an embarrassment factor and lack of aesthetic appeal as also influencing an individual's level of comfort [7].

This study demonstrates the importance of assistive devices for dressing. However, it does not specifically show the difference between upper and lower extremity dressing devices usage and needs. One study addressed this issue. The Use of and Self-Perceived Need for Assistive Device in Individuals with Disabilities in Taiwan

Table 1 Number and percentage of participants in four levels of functions in various types of basic ADLs [5]

ADLs	Able to perform Without ADs		Able to perform Occasionally use ADs		Able to perform Frequently use ADs		Unable to perform	
	N	%	N	%	N	%	N	%
Feeding	730	71.7	118	11.6	50	4.9	120	11.8
UE dressing	701	68.9	58	5.7	40	3.9	219	21.5
LE dressing	663	65.1	65	8.3	43	4.2	227	22.3
Grooming	665	67.3	93	9.1	103	10.1	137	13.5
Bathing	500	49.1	140	13.8	118	11.6	260	25.5
Toileting	563	55.3	109	10.7	114	11.2	232	22.6
Transferring	650	63.9	60	6.7	81	6.0	219	21.5
Walking	358	35.2	189	18.6	365	35.9	106	10.4
Communication	710	69.7	84	8.3	58	5.7	166	16.3

ADLs activities of daily living, *N* number, *ADs* assistive devices, *LIE* upper extremities, *LE* lower extremities
https://doi.org/10.1371/journal.pone.0152707.t002

investigated the degree of disability on the use of and self-perceived need for assistive devices. As shown in Table 1, a higher percentage of participants were able to perform feeding (71.7%), communication (69.7%) and upper extremities dressing (68.9%) activities without assistive devices. Walking and bathing were the activities that a relatively high percentage of participants were able to perform with the occasional or frequent use of assistive devices while bathing, toileting and lower extremities dressing were the activities that a higher percentage of participants were unable to perform and for which help from others was required.

The said study also examined the percentage of participants who used and needed assistive devices as well as 3 items with the highest percentage of participants who used and needed in each type of devices. It can be seen in Table 2 that there is 1.7% of participants who used the assistive devices for lower extremities but 7.9% of them perceived the need for these devices. This proves that there is a higher percentage of need of the lower extremity dressing aids as compared to its usage. To further understand the problems encountered in using a dressing aid for the lower extremities, another study explored the use of assistive devices for dressing by older people with impairments. Table 3 summarizes the results for reasons for dissatisfaction with dressing devices. Knowing these needs regarding lower extremity dressing aids and identifying the key factors on the dissatisfaction of the patients in using them are essential to design a device that is efficient, easy-to-use and one that meet their needs.

Market Analysis and Potential Users. We researched for commercially available products on lower extremity dressing devices but so far, there has not been any automated ones available in the market. We also managed to interview sellers of assistive devices regarding our initial idea on the automated lower garment dressing

Table 2 The percentage of participants who used and needed assistive devices [5]

Category of ADs AD items	Used (%)	Category of ADs AD items	Needed (%)
Feeding	16.2	**Feeding**	22.8
Lap board	5.2	Special utensils	8.6
Special utensils	4.4	Lap beard	7.1
ACs for grasp	1.7	Special plate/bowl	7.0
Dressing	12.9	**Dressing**	23.0
ADs for shoes wearing	4.2	ADs for shoes wearing	9.2
ADs for LE dressing	1.7	ADs for LE dressing	7.9
Special clothes	1.7	ADs for UE dressing	6.4
Grooming/bathing	25.4	**Grooming/bathing**	42.0
Bath bench	12.0	Bath bench	21.5
Anti-slip mats	7.6	Anti-slip mats	17.7
Lever handle faucet	5.6	Shower chair	8.6
Toileting	22.0	**Toileting**	28.9
Commode Chair	8.0	Commode chair	12.8
Bed pans and urinals	5.5	Raised toilet seat	7.0

Table 3 Reasons for dissatisfaction with dressing devices [6]

Reason not satisfied/used	Frequency	Percent of responses (%)
Don't need/use	56	65.1
Does not work well	14	16.3
Too difficult to use	11	12.8
Broken	2	2.3
Lost	2	2.3
Other	1	1.2

aid. Some responded that although it is a new idea, the patients might perceive it as complicated to use and prefer to pay someone to assist them in performing their activities of daily living.

However, most of the elderly and people with disability cannot afford to pay for someone to help them all the time. According to research shown in Fig. 1, most of elderly people live alone with 48.2% of the participants in this study of Elderly People's Use of and Attitudes towards Assistive Devices. That means there would be an increase to their needs of assistive devices to aid them on a daily basis. Currently, there are variety of assistive tools for dressing on the lower extremity. The most common ones are the sock and trouser dressing sticks. But usually, they have a single function such as putting on the garment but no function for taking it off. An

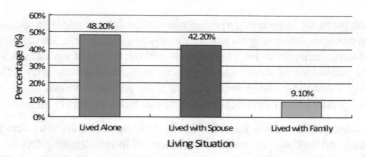

Fig. 1 Participants' living situations [8]

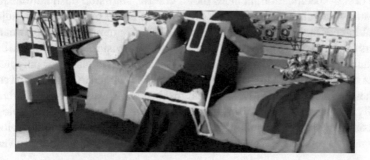

Fig. 2 Upper and lower garment dressing device [9]

existing dressing device for both the upper and lower garment is shown in Fig. 2. The design is efficient and easy to use. However, the users may encounter difficulties in bending their bodies especially when they have limited shoulder movement.

There are various reasons why dressing and undressing becomes difficult for an individual. Some of these are due to back pain and inability to lean forward, one sided muscle weakness after a stroke, unbalanced and poor coordination of the body, and many others. Nonetheless, previous studies showed a higher percentage of need for lower extremity dressing devices for elder people who are either or both frail and suffer with certain disabilities. Response from some sellers of assistive devices for elders, interviewed by the research team, commented that making an automated dressing device might be too much of a hassle especially for elder people. They mentioned that the elders will prefer to pay someone who will take care of them. However, the elderly people do not only lack the means to pay someone but there is also an increasing number of elderly needing assistance. According to the World Health Organization, between 2015 and 2050, the proportion of the world's population over 60 years will nearly double from 12 to 22%. The pace of population ageing is much faster than in the past. By 2020, they projected that the number of people aged 60 years and older will outnumber children younger than 5 years. Moreover, in 2050, 80% of older people will be living in low- and middle-income countries.

Furthermore, we interviewed a person who suffers from hemiplegia and one of his dressing problems is reaching his pants with only one healthy arm. He mentioned that it was always a struggle to dress and undress without the aid of a person or a device. With the prevalence of stroke and increasing number of people who suffer from monoplegia; and the growing population of elder people, the need for an efficient and cost-friendly lower extremity dressing device will most likely increase.

Design Concept. To address the difficulty in dressing and undressing activity such as reaching and bending, we designed an automated lower extremity dressing device that is also portable enough to be moved around. Figure 3 shows the sketch of the device.

The design specification includes motor control, start/stop push buttons for each side, battery and charger, pulleys, stopper wheels at the base, base bar, adjustable garment hanger and the additional support. Table 4 summarizes the part name and numbers for the design. For the mechanical parts, we envisioned the design to be made up of stainless steel to prevent corrosion and oxidation, and plastic or polypropylene vinyl for cost-reduction. Most of the parts will also be painted in a minimalist color for better aesthetics. Subsequently, electrical mechanism is also added to minimize the difficulty of mechanical control that is present in the existing assistive technology. The motor control will drive the pulley system to rotate in a clockwise or counterclockwise direction. This rotation is what drives the garment hanger system to move vertically

Fig. 3 Design sketch of the automated lower extremity dressing device

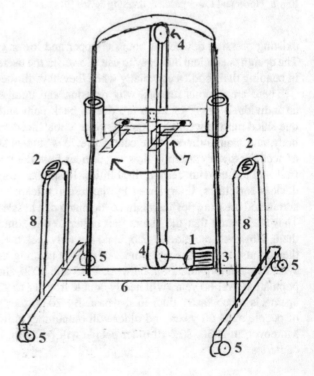

Part number	Part name
Table 4 Summary of part numbers and names	
1	Motor control
2	Push buttons
3	Battery and charger
4	Pulleys
5	Stopper wheels
6	Base bar
7	Garment hanger
8	Additional support

for automated dressing and undressing of the lower garment. Two push buttons will be placed on both sides that are connected in parallel with each other to start or stop the circuit. One button is connected in such a way that the motor rotates in one direction and the other button for the reverse direction. Once they are released, there will be no current flowing the circuit and thus, the system stops moving. The whole circuit is powered by a battery. Moreover, to accommodate different sizes of pants, we designed the garment hanger system to be adjustable horizontally. The additional support on the side and the wheels make the whole device portable enough to be moved around. In this design, the patient must fold the waist band of the pants to the garment hanger and step inside the pants when it is lowered to the desired level. After that, the garment hanger can be moved vertically by pushing the button on the side.

3   Methods

3.1   Mechanical Simulation on SolidWorks

For the mechanical simulation of the concept, we used SolidWorks software to have a thorough design of the parts with the dimensions seen in Fig. 4. The first part that was being made is the foundation and base which included the 4 wheels with stopper as shown in Fig. 5. The side support can act as a cane which can help the user to stand and walk.

The architectural diagram of the next part is shown in Fig. 6. It displays the pulley mount on the center top and bottom of the device. This is composed of a pulley shown on the left of Fig. 7 and a clamp of the fixed pulley on the right of Fig. 7. This is one of the core elements that drive the whole system. The yellow push buttons located on the hand support can also be seen in Fig. 9. The two sets of push buttons basically have the same parallel functions. When either is pressed, it will make the motor rotate in a certain direction. When released, the circuit is open circuit which stops the motor from rotating. Two sets are made in order for the user to have the

Fig. 4 Lower extremity
dressing device dimension

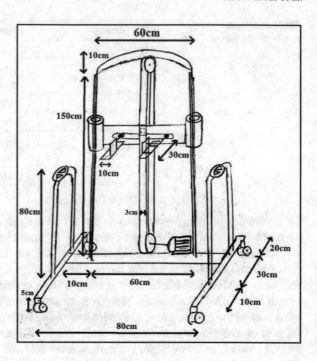

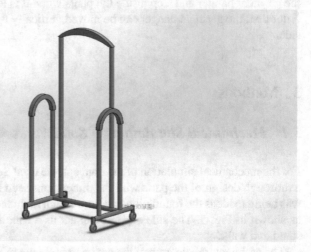

Fig. 5 Design foundation and base

convenience in using either of the two for monoplegic patients who have paralyzed right or left hand.

Figure 8 exhibits the different perspective of the design architecture with the added belt on the center that connects the 2 pulleys and make up the whole pulley system. The rotation of the pulley drives the belt to move vertically.

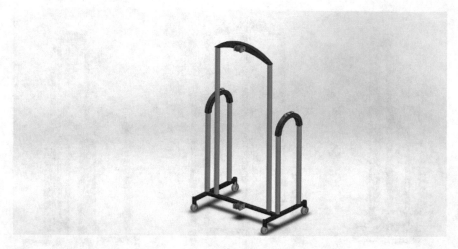

Fig. 6 Design foundation and base with pulley mount

Fig. 7 (Left) pulley and (right) clamp

The key element of the design is the garment hanger viewed in different perspective shown in Fig. 9. This is where the user can fold the waist band of the pants. This part is connected to the belt by attaching the belt between its layers. Through this connection, whenever the belt moves vertically, the garment hanger also moves along with it.

The center design of this part is meant to be adjustable to accommodate the different sizes of waist and for more convenient user experience. The hollows on the left and right sides are combined with the tubular sides of the center foundation in Fig. 10. Finally, the whole design on different angles is shown in Fig. 11. The simulation ran smoothly in SolidWorks. Since we could not add the motor control circuit in the SolidWorks simulation, we just set pulley to rotate in both clockwise and

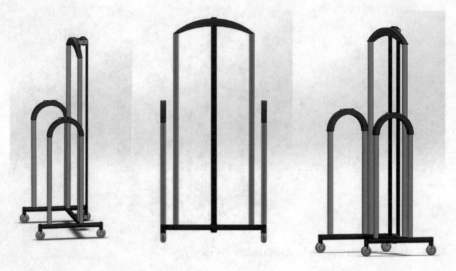

Fig. 8 Design foundation with belt

Fig. 9 Garment hanger design with different views

Fig. 10 Attachment of the
garment hanger to the base

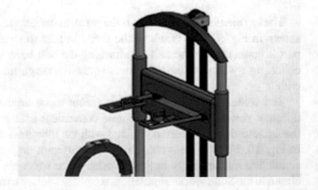

Fig. 11 Full design of the lower extremity dressing aid on different views

counterclockwise direction. Through this, the belt moves with the pulley making the garment hanger moves with it vertically. This movement is precisely what we need in our design to accommodate both dressing and undressing of the lower garment.

Electrical Simulation on Proteus. For the electrical mechanism of the device, we simulated it on Proteus 8 Professional. This study used L293D motor driver to drive the servo motor. Figure 12 shows the circuit diagram. It has 2 push buttons that are connected to the first and second input of L293D. Outputs 1 and 2 are connected to the servo motor. When the first button is pressed, the first input is enabled which will then make the output 1 high. This will result to the rotation of the motor in counterclockwise direction. On the other hand, when the second button is pressed, the motor will rotate in clockwise direction. Push buttons are used instead of a toggle switch to make it easier for the user to stop the system when desired. Once the buttons are released, there will be no current flowing through the circuit and therefore, making the whole system stop.

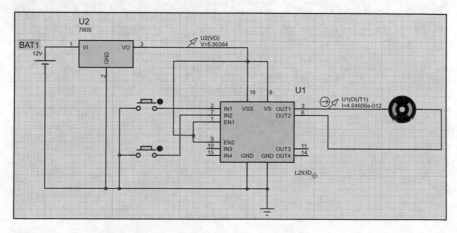

Fig. 12 Circuit diagram for motor control

The whole system is powered by a 12 V battery. Although, in the simulation it was not directly connected to the motor because of the default state of Proteus to have the motor powered by an internal source, we assumed that in actual experiment, the battery will be connected to it. This same battery is also regulated to output 5 V via 7805 voltage regulator to power the L293D motor driver. We decided to use only one battery to lower both the cost and weight of the whole device.

However, one of the challenges encountered is how to properly define the speed of the motor. Since the speed of the motor is dependent on the value of the current through it, we needed to find a way to manipulate the current flowing through the motor. To do that, we simulated the circuit by having different supply voltages for the L293D motor driver.

Figure 13 shows that when the supply voltage is 5 V, the current through the motor will be 357 mA. Adjusting the supply voltage to a higher value such as 9 V will also increase the current to 643 mA as seen in Fig. 14. Therefore, we only need to control the supply voltage to adjust to the desired output current as long as that value is within the operational voltage range of the motor driver.

For the purpose of this study, we assumed that the 5 V supply for the motor driver by regulating the 12 V battery will be sufficient enough since we wanted the motor to rotate with average speed. Thus, it is evident that the circuit is flexible enough to cater the desired output current through the motor which ultimately drives the vertical movement of the garment hanger system of the whole device.

Component Specifications. The L293D motor drive that outputs the current for the motor to rotate has a wide supply-voltage range that is within 4.5–36 V based on Texas Instruments' L293x data sheet. This means if further research suggests to increase the speed of the motor then it will be easy to adjust the supply voltage of the motor driver within this range. This motor driver is specifically good for DC

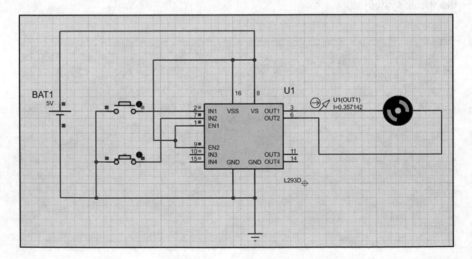

Fig. 13 Circuit simulation with 5 V supply

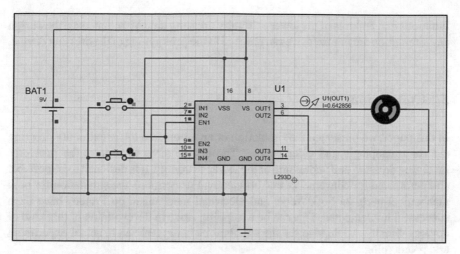

Fig. 14 Circuit simulation with 9 V supply

motor drivers. The motor that will be used for this device is a metal DC geared motor with encoder. It is known to have a no-load current of 350 mA. That means it can operate with at least 350 mA of current through it. Based on the previously mentioned simulation, the current through the motor when we have a 5 V motor driver supply voltage is 357 mA that will be sufficient to rotate the motor. In addition, it has a stall torque of 18 kg cm this means that if we connect a load of 1 cm from the center of the shaft, it can lift a maximum load of 18 kg. Since the garment hanger along with the pulley system will weigh less than 18 kg, the motor will sustain the weight of the system. Moreover, the motor itself has a weight of 205 g which is an advantage for the design because we need to ensure that the parts are not that heavy.

4 Discussion

4.1 Limitation

The lower extremity dressing aid is limited only to vertical movement—upwards for dressing and downwards for undressing of the lower garment. Since we decided to make it portable, it must use a battery as a power supply. Therefore, the battery should be rechargeable to lower the cost of maintenance. The height of the device can also be improved by considering the physical dimensions and social characteristics of the home where the user lives.

Future Work. For further research, we can look more into the analysis of the target market's perception of this kind of device and their specific needs regarding the functions and features of the lower extremity dressing aid. After that, once the design and simulation are finalized, we can proceed to the prototyping of the dressing device.

5 Conclusion

In this paper, we presented the simulation of an automated lower extremity dressing aid for monoplegic and elder people. The device aims to address the problems involving dressing and undressing of the lower extremity. It is designed to be portable and automated for convenience and efficiency. Furthermore, the simulation for both mechanical parts on SolidWorks and electrical mechanism on Proteus have been successful implemented. This can be a stepping stone in the evolution of this kind of devices. Finally, we hope to see the simulations being realized in actual experiment and prototyping.

References

1. Wade, D., Hewer, R.: Functional Abilities After Stroke: Measurement, National History and Prognosis, vol. 50, pp. 177–182 (1987)
2. Park, J., Cho, J., Choi, J.: Design and development of a foot position tracking device. In: 30th Annual International Conference of the IEEE, pp. 2903–4906. IEEE, Canada (2008)
3. Fenichel, G.: Clinical Pediatric Neurology, 6th edn. Saunders, Philadelphia (2009)
4. Gitlin, L., Levine, R., Geiger, C.: Adaptive device use by older adults with mixed disabilities. Arch. Phys. Med. Rehabil. **74** (1993)
5. Yeung, K., Lin, C., Teng, Y., Chen, F., Lou, C., Chen, L.: Use of and Self-Perceived Need for Assistive Devices in Individuals with Disabilities in Taiwan (2016)
6. Mann, W.: Problems with dressing in the frail elderly. Am. J. Occup. Ther. **59**, 398–408
7. Bruno, R.: Psychologic predictors of pain management program outcome. Arch. Phys. Med. Rehabil. (1990)
8. Yeh, H.: Elderly People's Use of and Attitude Towards Assistive Device, unpublished
9. Sheri D. Bean Co., Miracle Dressing Aid-™, Web: https://www.youtube.com/watch?v=pb9Ovt BD0O0. Last accessed 2019/06/21

Obstacle Detector and Qibla Finder for Visually Impaired Muslim Community

Kristine Mae Paboreal Dunque, Aaron Raymond See(iD), Dwi Sudarno Putra, Rong Da Lin, and Bo-Yi Li

1 Introduction

On the report of the World Health Organization's "Blindness and Vision Impairment Fact Sheets", updated in October 2019, an estimate of 2.2 billion individuals have vision impairment or blindness [1]. There could be 115.0 million persons who will go completely blind by 2050, as shown in Fig. 1, up from 38.5 million in 2020 [2]. Children below age 15, and people aged 50 and over, are at risk of acquiring vision impairment and blindness. Approximately 81.0% of all people who are blind or have moderate to severe vision impairment are in the elderly age bracket.

The results from the "Global Vision Database" and meta-analysis showed that the majority of blind people reside in South Asia (11.7 million), East Asia (6.2 million), and Southeast Asia (3.5 million). The burden of eye conditions and vision impairment remains a serious health problem in developing countries [3] and is often greater in people who are living in rural areas, those with low income, the older people, the ethnic minorities, and the indigenous populations. Indonesia has the highest blindness rate in the Southeast Asian region. The majority of the 3-million, visually-impaired population have Islamic faith [4, 5].

Salah is the second pillar of the Islamic faith. It is a form of worship made up of Rak'ah or units or prayer. Every Rak'ah has the same basic steps illustrated by Masjid ar-Rahmah in Fig. 2, within it, includes a part when the believer bows down

K. M. P. Dunque · A. R. See (✉) · D. S. Putra · R. Da Lin · B.-Y. Li
Southern Taiwan University of Science and Technology, Tainan 71005, Taiwan (Republic of China)
e-mail: aaronsee@stust.edu.tw

K. M. P. Dunque
University of Science and Technology of Southern Philippines, CDO Campus, 9000 Cagayan de Oro City, Philippines

D. S. Putra
Universitas Negeri Padang, Padang 25171, Indonesia

© Springer Nature Switzerland AG 2021
C. T. Lim et al. (eds.), *17th International Conference on Biomedical Engineering*,
IFMBE Proceedings 79, https://doi.org/10.1007/978-3-030-62045-5_6

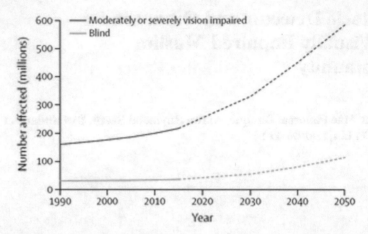

Fig. 1 Global trends and predictions of numbers of people who are blind or moderately and severely vision impaired, from 1990 to 2050

Fig. 2 Ruku (left image) and sujud (right image)

(make ruku) and prostates (make sujud). An important condition for the salah is for the person to face the direction of Qibla, precisely oriented towards the Kaaba in the city of Mecca, Saudi Arabia. The visually impaired Muslim community will have difficulty determining the Qibla and thus may restrict their ability to perform their prayer duty. Also, they need to protect themselves from hitting objects or people during their worship.

Subsequently, a great deal of study has been carried out on wearable devices for the visually impaired population but only a few numbers of assistive technologies are specifically tailored for the Muslim blind society. Salleh et al. [6] developed a low-cost, easy-to-handle device to detect the Qibla direction using a magnetic compass, sensor assembly, and buzzer. While Mutiara and Sanjaya et al. [7, 8] used Arduino microcontroller in designing their prototype system to detect the obstacle, to give the Qibla direction and to display the prayer time. And Asrin et al. [9] incorporated a voice command function into a cane in determining the Qibla. Some of the existing devices are handheld which require constant hand interaction. In the

Fig. 3 Indonesian Peci (left image) and aviation hijab (right image)

above-mentioned method, the blind person may lose a grip on the device. Others use a power bank on the hardware which may add a little weight on the portable device.

The purpose of this study is to develop a hands-free, lightweight, and wearable assistive device that can be embedded in the commonly used headwear worn among the Muslims. All the electronic components will be mounted and fixed inside the Peci for men and Hijab for women, as shown in Fig.3, providing safety, accuracy, and comfort for the end-user. Then a mobile application will be integrated into the whole system. Therefore, besides augmenting the user capability of securing the correct prayer direction, this study will also be protecting the visually impaired individual from hitting objects or people during their salah daily prayers. We will also take advantage of the global positioning system or GPS, and google map. These digital mapping technologies will aid in detecting the current location of the blind individual from the mobile application of the family member's smartphone. The obstacle detector, Qibla finder, and mobile application will guarantee Muslim blind society to be aware of their surrounding during prayer worship. After the introduction in Sect. 1. The remainder of this paper is organized as follows. Section 2 presents the materials and methods. Experimental results are validated in Sect. 3. Discussion and implementation are proposed in Sect. 4. Finally, Sect. 5 summarizes the conclusions and future work.

2 Methodology

2.1 Materials

The proposed system is shown in Figs. 4 and 5, that demonstrate the system architecture with its four main parts, namely, input, processor, output, and power supply.

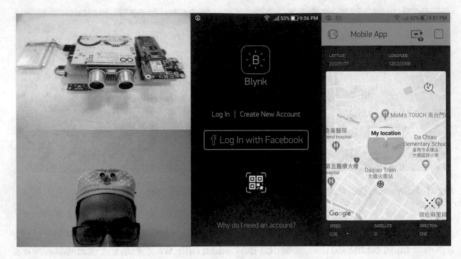

Fig. 4 Proposed system

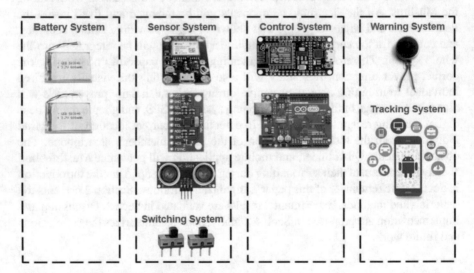

Fig. 5 System architecture

Input (Sensor and Switching System). The system is endowed with three types of sensors, namely, the NEO-6M GPS module, MPU9250 digital compass, and HC-SR04 ultrasonic sensor. The GPS receiver provides continuous real-time, 3-dimensional positioning, navigation, and timing based on the trilateration mathematical principle. The digital compass coupled with the trigonometry spherical triangle formula determines the Qibla at the current location of the blind Muslim person. The proposed method uses the difference, between the locations of the satellites in space and the blind user, and the direction of the magnetic field where the wearer of the system is facing, to secure the correct Qibla direction. The ultrasonic sensor comprises one transmitter and one receiver that can measure the distance of an object from 2 to 400 cm using ultrasonic sound waves, and relays back information about the object's proximity. The system will detect the presence of an object's location within 50 cm in front of the ultrasonic sensor mounted on a Peci or Hijab. If an obstacle is recognized such as a wall or a person, the vibration motor will be triggered, protecting the visually impaired Muslim during worship prayer. Control will be done through the use of two slider switches for the power and the GPS functions.

Processor (Control System). Furthermore, the whole architecture is managed by Arduino Uno and ESP8266 microcontrollers. The Arduino Uno sets up communication among the ultrasonic, magnetometer, GPS, switches, and motor. Codes written here will be responsible for the obstacle detection and the calculation of the precise Qibla direction. We employed and coded ESP8266 for its Wi-Fi feature, which will be used for sending the accurate location of the blind individual. The GPS module and mobile application are connected to this microprocessor that will supply the latitude and longitude coordinates of the blind individual. In this way, the mobile application developed through the Blynk, an IoT platform, will be able to navigate the family member through the map widget on their smartphone. Opening the app will then show the exact map location with the corresponding coordinates of their blind loved one.

Output (Warning and Tracking System). A precision 10 mm brushless vibration motor 3-mm type is used as a non-audible indicator. Either an obstacle is detected or a Qibla direction is located, this tiny button-type haptic alarm with a 0.4 G low vibration force will move, which will be easily felt by the blind consumer. It will guarantee the visually impaired user can avoid hitting their heads on hard objects while performing the Rak'ah. And for the tracker feature of the system architecture, a mobile application is developed, to display the latitude, longitude, and map location of the blind individual. The Blynk platform is simple to operate and can run on over 400 hardware modules, visualize and plot data from any sensor, get push notifications, send emails, and control actuators.

Power Supply (Battery System). It is powered by two small-scale 3.7 V lithium polymer rechargeable batteries, which are suitable for lightweight applications and can deliver stable voltage over a long distance.

2.2 Methods

Altium Designer was used for sketching the schematic diagram and the printed circuit board layout. LPKF ProtoMat was used to print the module as shown in the system assembly of Fig. 6.

The wearable device has two slider switches as shown in the schematic diagram of Fig. 7. One functions as the power switch for the activation of the battery supply

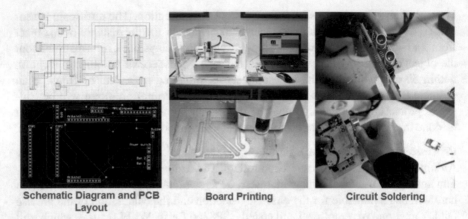

Schematic Diagram and PCB   **Board Printing**   **Circuit Soldering**
            Layout

Fig. 6 System assembly

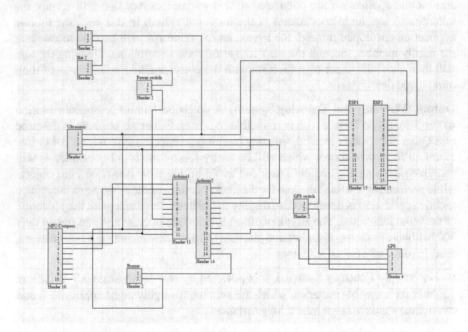

Fig. 7 Schematic diagram

and obstacle detection. While the other functions as the GPS switch for the operation
of the Qibla determination and mobile application. When the power switch is turned
on, the ultrasonic sensor is then activated. The Trig and Echo are both I/O pins and
are connected to the I/O digital D3 and D2 (pins 4 and 3 from Arduino2, Header
14) terminals of the Arduino Uno. The transmitter sends a signal. Once an object
is detected, the echo pulse will be reflected at the receiver. The time between the
transmission and reception of the signal allows us to calculate the distance of an

object. Once the object is within the 50 cm allowable range, the motor will vibrate indicating an obstacle. The sensor is most accurate when the object to be detected is directly in front of it. The testing recommends a 20° angle of operation window for exact readings. The accuracy to detect objects degrades as the angle of coverage increases. When the second switch is triggered, the Qibla finder mode will be initialized. The SCL and SDA pins of the digital compass are connected to the analog A5 and A4 terminals (pins 1 and 2 from Arduino1, Header 11) of the Arduino Uno, the component responsible to get the current compass heading. The RX and TX pins of the GPS receiver are connected both to the digital D9 and D8 (pins 11 and 10 from Arduino2, Header 14) terminals of the Arduino and to the digital D1 and D2 (pins 1 and 2 from ESP1, Header 15) terminals of the ESP8266 Wi-Fi Microchip. The configuration will provide the longitude and latitude data of the blind individual. As long as the values of the compass header and the Qibla direction are matched, the vibration motor which is connected to D12 (pin 14 from Arduino2, Header 14) will then be activated. The user needs to turn around to locate the correct Qibla direction. The corresponding device coordinates will be sent to the Blynk mapping widget of the family members, monitoring the real-time location of the blind loved one. During the testing, the device should be near the window or operated in an open space for the GPS to work properly. The equivalent PCB layout is shown in Fig. 8.

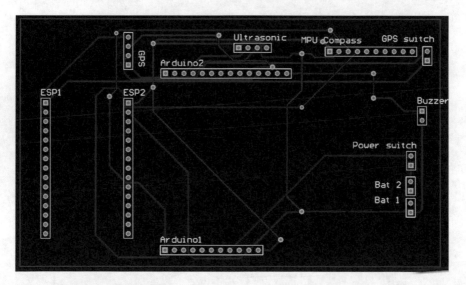

Fig. 8 Printed circuit board layout

3 Experimental Results

3.1 Software

The working principle of the prototype is illustrated in Fig. 9 flowchart. The entire design architecture has 3 objectives for obstacle detection, Qibla determination, and mobile application. The system has two switches for the power and GPS operations. With the switch 1 mode, the supply and obstacle detector functions are activated. With the switch 2 mode, the GPS is triggered, initializing the Qibla finder and mobile application.

Objective 1. The transmitter sends a high-level signal of 8 cycles of ultrasonic burst at 40 kHz and when the said signal detects an object, it is then bounced off at the receiver. The time between the transmission and reception of the signal reflection allows us to calculate the distance to an object. During the detection of an obstacle, the time it takes for the reflected waves to be bounced back, which is proportional to the distance, will be sent to the Arduino Uno. Eventually, the output port will trigger the motor to vibrate, signaling that the ultrasonic sensor in front recognizes an object that is within the 50 cm range allowance.

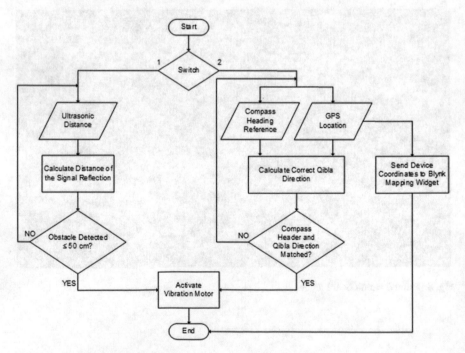

Fig. 9 System software algorithm

Objective 2. The GPS module will provide the location coordinates such as the latitude and longitude of the carrier's device. The information will be processed along with the direction obtained from the digital compass. The value will be compared with the exact position coordinates of the Kaaba direction in the city of Mecca, Saudi Arabia using the input formulas operated by the Arduino Uno. Consequently, the vibration motor will be activated as soon as the blind user turns to the correct Qibla direction, that is when the compass header matches the Qibla direction. We applied the formula used by [8] to determine the accurate Qibla direction:

$$\tan(Q) = \frac{\sin(\lambda_L - \lambda_M)}{cos\varphi_L.tan\varphi_M - sin\varphi_L.\cos(\lambda_L - \lambda_M)} \tag{1}$$

where

$$\varphi_M = Makkah(Qibla)Latitude = 21°25'21.21'' \tag{2}$$

$$\lambda_M = Makkah(Qibla)Longitude = 39°49'34.56'' \tag{3}$$

$$\varphi_L = Latitude of Current Location \tag{4}$$

$$\lambda_L = Longitude of Current Location \tag{5}$$

Objective 3. The ESP8266, GPS, and map widget developed in Blynk will track the position of the assistive device worn by the visually impaired person. In this way, the family member can monitor and will be notified of the real-time location of the blind loved one with the aid of their smartphones.

3.2 Hardware

The electronic components are all soldered in a single printed circuit board as illustrated in Fig. 10. The system design will eliminate the unnecessary device placement on the forehead, hands, and waist, which make wearable aids inconvenient for the person with an impairment. It will use a single, comfortable, and lightweight headwear. All the navigation aids are compactly placed as one. It will facilitate mobility without carrying a lot of assistive device technologies. Blind people rely on hearing environmental cues for key tasks, such as awareness, orientation, mobility, and safety. As a result of consideration, the design will use vibration motors instead of buzzers, that can be easily felt by the blind user once the obstacle or Qibla is detected. The prototype will use ultrasonic distance sensors optimally positioned in front of the Indonesian Peci, as demonstrated in Fig. 11. The orientation will protect the head

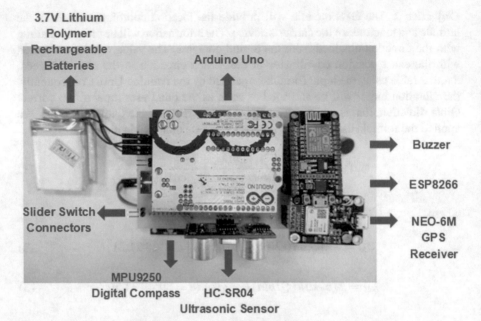

3.7V Lithium Polymer Rechargeable Batteries

Arduino Uno

Buzzer

ESP8266

Slider Switch Connectors

NEO-6M GPS Receiver

MPU9250 Digital Compass

HC-SR04 Ultrasonic Sensor

Fig. 10 System hardware design

Fig. 11 Obstacle detector and qibla finder for visually impaired muslim community prototype embedded in a traditional Indonesian headwear peci

from hitting hard objects. The slider switches are located at the back of the traditional Muslim headwear for easy access. And finally, the prototype board will be fitted inside at the upmost part of the headwear. This compact device is efficiently fastened just like an item of their everyday clothing.

Obstacle Detector. The first image displays the working program of the distance sensor in an Arduino IDE environment. After the prototype testing against the wall is performed as portrayed in the second image, the device is installed inside the upmost part of the Indonesian Peci. The assistive device can perform well in determining an object in front of it as shown in the third image of Fig. 12.

Fig. 12 Prototype board and embedded device testing: obstacle detection

The ultrasonic sensor is most accurate when the object to be detected is directly in front of it. The researchers decided to examine 10 distance trials and 3 sets of angles to check the optimal radial coverage of the distance sensor. For the experimental setup shown in Fig. 13, the team used a single ultrasonic sensor, one Arduino Uno microcontroller, a push–pull ruler, a test object which has a 9 cm diameter, and a crafted paper compass. From the results achieved in Figs. 14, 15, and 16, it can detect obstacles clearly from 0° to 20°. The IBM SPSS Statistics Editor and EXCEL are used to plot the results.

Qibla Finder. The programmed Qibla calculation along with the latitude and longitude inputs from the GPS receiver, and the current heading direction of the digital compass is run, calibrated, and tested in Arduino IDE as illustrated in the first image of Fig. 17. Then we use an android mobile application called 100% Qibla Finder, as a reference compass to verify the accuracy of the calculated Qibla as shown in the second image. It will be challenging for the blind individual to use this App since they need to see the arrow on the map pointed to the direction of the Kaaba in Mecca. Hence they will have difficulty in adjusting their direction before they can start their prayer. With our system as demonstrated in the third image, the blind person just

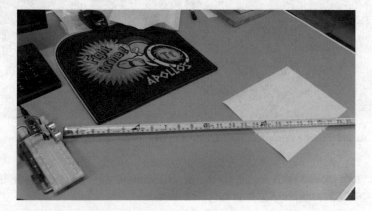

Fig. 13 Actual distance measurement testing

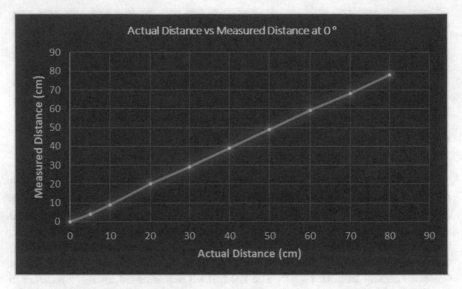

Fig. 14 Actual distance versus measured distance at 0°

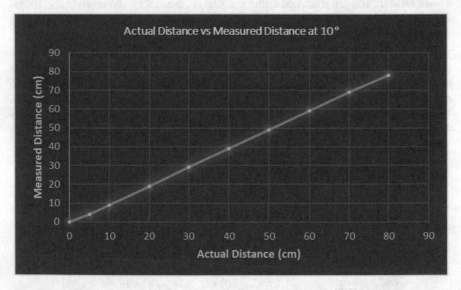

Fig. 15 Actual distance versus measured distance at 10°

needs to turn around until the haptic motor will vibrate, signaling that the individual is now facing the correct Qibla prayer direction.

Mobile Application. For the developed mobile application in Fig. 18, the current location of the visually impaired individual can be accessed through the web and smartphone, as represented in the second and third images respectively. Once the

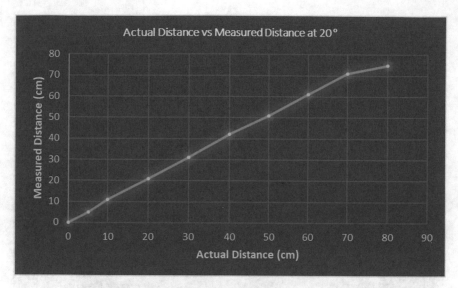

Fig. 16 Actual distance versus measured distance at 20°

Fig. 17 Prototype board and embedded device testing: qibla finder

GPS switch is turned on, the wearable device will be connected to the Wi-Fi of the ESP8266, subsequently, the server will get started as shown in the first image. The web option will present the latitude, longitude, date, and time. As a family member, you have the option to view the exact location of your blind loved one on the Google map by clicking a part of the webpage. The Blynk map widget will also give the latitude and longitude of the current location of the blind loved one. Moreover, the mobile application from your smartphone will also display the exact location of the blind individual on the Google map. Since almost everyone cannot leave their homes without bringing their smartphones, having a mobile application that can track and monitor a blind member of their family can be very handy and convenient.

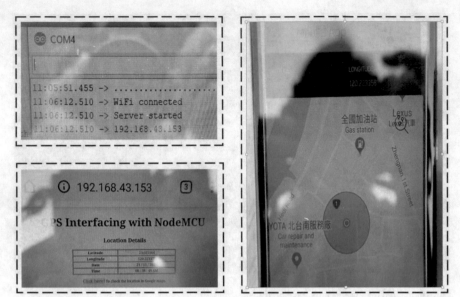

Fig. 18 Prototype board and embedded device testing: mobile application

4 Discussion and Implementation

The development of the obstacle detector has been validated. It ensures the safety of visually impaired Muslim society during their worship prayer. In this paper, the Qibla finder has been calibrated and tested for accuracy. The mobile application has successfully provided the precise location of the blind person. In order to get real feedback from the target end-users, beta testing will be employed in an additional experiment. Identified visually impaired Muslims will try the device. It will be performed during prayer worship, five times a day in the individual's location. In this way, we can guarantee that the product is ready for the intended users. Feedback and suggestions will be collected and evaluated based on the participant's experience with the proposed technology.

In this section, we also present a comparative survey between the assistive devices available in the market and the proposed system for Muslim visually impaired individuals. All wearable devices that are used by blind people helping them daily, making their activities easier and providing safe mobility are condensed in Table 1. From the presentation of the different studies, the weight of the assistive devices introduces a drawback, which hinders easy mobility among the visually impaired individuals. Some assistive devices, that are worn on the head use a safety helmet that approximately weighs 400 g or more. In designing the assistive device, we take into account the lightweight property of an Indonesian Peci, which is mostly made from polyester fabric and usually weighs from 86 to 200 g only. Furthermore, other existing assistive prototypes are handheld which requires the users to hold and grip

Table 1 Existing wearable technologies for the blind worn on the head and available portable assistive devices for the Muslim blind community

Assistive device	System architecture	Working principle
	A head-mounted device equipped with 2 cameras that send visual information to a portable computer used for video-processing. The head tracking system is an Xsens Mti composed of 3-axis accelerometers, a magnetic compass, and gyroscopes [10]	An assistive device for the blind based on adapted GIS, and fusion of GPS with vision-based positioning
	The system utilized a Microsoft Kinect sensor, a vibrotactile waistbelt built with Arduino LilyPad vibe boards and a simple backpack construction that carried the laptop. They also used a 12 V battery pack and a Bluetooth headset [11]	A mobile navigational aid that uses the Microsoft Kinect and optical marker tracking to help visually impaired people find their way inside buildings
	The device used a magnetic compass and buzzer. It also integrated a LED indicator to notify the non-blind user of the exact prayer direction of Qibla. It is then powered by a 23.0 A size battery [6]	The development of a low-cost and easy-to-handle device to detect the Qibla direction for the visually impaired Muslims
	The prototype used an ultrasonic distance PING sensor, electronic compass CMPS 11 sensor module, Arduino UNO, DFPlayer mini integrated serial MP3 module, pushbutton, led indicator, and buzzer. All the navigational components are attached to the traditional white cane [7]	The study designed a prototype cane that can detect obstacles, holes and give the correct Qibla prayer direction

(continued)

Table 1 (continued)

Assistive device	System architecture	Working principle
	The embedded device employed NEO6MV2 GPS module, HMC5883L compass module, 16 × 2 size LCD display, buzzer, and Arduino Uno board. The entire architecture is sustained by a power bank [8]	In this research, a portable assistive device is developed to display the real-time Qibla direction and Sholat prayer times
	The handheld prototype used a Raspberry Pi 3 model B as a processor, HMC5883L as a compass sensor, and a microphone as voice command input [9]. The corresponding voice responses are transferred via the headphones. The whole system architecture is supplied by a power bank	An interactive function cane is developed that will enable the blind user to determine the exact direction of the Qibla during prayer worship by voice command

the device all the time. The proposed solution offers a discreet and hands-free usage of the obstacle detector and Qibla finder. Although a lot of the mobile applications are now available to instantly find the direction of Qibla from anywhere of the world and perform prayer accurately, still only a small number has accessibility support for the visually impaired people through voice-over. And right now no mobile application has been developed, that can send the exact mapping coordinates of the blind person enabling the family members to track and can monitor their visually impaired loved one. The Blynk developed mobile application will promote the reliable and independent mobility of the Muslim community.

5   Conclusion and Future Works

A portable, lightweight, and low-cost device was successfully designed and developed to assist visually impaired Muslims. With this wearable system, the Muslim blind community can independently perform prayers anytime and anywhere that is safe and accurate. The future work for this research entails making the module more compact and small-scale.

References

1. WHO Fact Sheets: https://www.who.int/news-room/fact-sheets/detail/blindness-and-visual-impairment. Last accessed 2020/02/15
2. Bourne, R.R.A., Flaxman, S.R., Braithwaite, T., et al.: Magnitude, temporal trends, and projections of the global prevalence of blindness and distance and near vision impairment: a systematic review and meta-analysis. Lancet Glob. Health **5**(9), 1–10 (2017)
3. Jaggernath, J., Øverland, L., Ramson, P., et al.: Poverty and eye health. Health **6**(14), 1849–1869 (2014)
4. UNDP Indonesia: https://www.id.undp.org/content/indonesia/en/home/presscenter/articles/2017/08/31/_leave-no-one-behind--undp-aims-to-champion-the-rights-of-visual.html. Last accessed 2020/02/15
5. IAPB Indonesia: https://www.iapb.org/wp-content/uploads/VIsion-2020-workshop-2014-Report-Indonesia.pdf. Last accessed 2020/02/15
6. Salleh, I., Zain, M.H., Soad, M.H., et al.: A qibla compass for visually impaired muslims. J. Telecommun. Electron. Comput. Eng. **8**(7), 13–16 (2016)
7. Mutiara, G.A., Hapsari, G.I., Rijalul, R.: Smart guide extension for blind cane. In: The 4th International Conference on Information and Communication Technology, pp. 1–6. IEEE, Bandung, Indonesia (2016)
8. Sanjaya, W.S.M., Anggraeni D., Nurrahman F.I. et al.: Qibla finder and sholat time based on digital compass, gps, and microprocessor. In: The 2nd Annual Applied Science and Engineering Conference, pp. 1–9. IOP Publishing Ltd, Bandung, Indonesia (2018)
9. Asrin, A., Hapsari, G.I., and Mutiara, G.A.: Development of qibla direction cane for blind using interactive voice command. In: The 6th International Conference on Information and Communication Technology, pp. 216–221. IEEE, Bandung, Indonesia (2018)
10. Brilhault, A., Kammoun, S., Gutierrez, O., Truillet, P., Jouffrais, C.: Fusion of artificial vision and gps to improve blind pedestrian positioning. In: The 4th IFIP International Conference on New Technologies, Mobility, and Security, pp. 1–5. IEEE, Paris, France (2011)
11. Zöllner, M., Huber, S., Jetter, H.C., Reiterer, H.: NAVI-a proof-of-concept of a mobile navigational aid for visually impaired based on the microsoft kinect. In: Campos, P., Graham, N., Jorge, J., Nunes N., Palanque, P., Winckler, M. (eds.) Human-Computer Interaction 2011, INTERACT, vol. 6949. Springer, Berlin, Heidelberg (2011)

Rejecting Artifacts Based on Identification of Optimal Independent Components in an Electroencephalogram During Cognitive Tasks

K. Kato, K. Suzuki, T. Suzuki, and H. Kadokura

1 Introduction

An electroencephalogram (EEG) is often contaminated by eye-blink and eye move-ment artifacts. This results in spurious neural activities in the EEG data incorrectly appearing as event-related potential (ERP) components. To address this problem, independent component analysis (ICA) can be employed for the rejection of artifacts in EEGs [1–6]. In a previous study, we proposed an identification technique for the optimal ICs of electrical noise generated by artifacts, where we integrated ICA and K-means, which is a clustering algorithm used in machine learning [7]. However, we evaluated the performance of the method for only artificial EEG noise superimposed on an eye-blink and eye movement template on a resting EEG. There is insufficient evidence for the practical use of this method. Therefore, in this study, we evaluated the performance of the proposed method by using real EEG data during an implicit association task (IAT), which enabled the detection of implicit biases in individual subjects.

2 Cognitive Task and EEG Measurement

We employed a cognitive task, namely an IAT [8–10], that enables the detection of implicit biases in individuals. During the task, images or characters were presented visually at an angle of $16° \times 11°$. The EEG was contaminated with artifacts generated by blinking and eye movement. The task comprised seven blocks, namely, blocks 1–7. The EEGs were measured for blocks 3, 4, 6, and 7, consisting of 20, 40, 20, and 40 stimulus presentations, respectively. The remaining blocks were practice blocks.

K. Kato (✉) · K. Suzuki · T. Suzuki · H. Kadokura
Tohoku Gakuin University, Tagajyo, Miyagi, Japan
e-mail: k_kato@mail.tohoku-gakuin.ac.jp

© Springer Nature Switzerland AG 2021
C. T. Lim et al. (eds.), *17th International Conference on Biomedical Engineering*,
IFMBE Proceedings 79, https://doi.org/10.1007/978-3-030-62045-5_7

A fixation cue, and a stimulus consisting of an image or a character were presented for 1000 and 1500 ms, alternately.

The EEG signal was acquired at F_{p1}, F_{p2}, F_7, F_3, F_z, F_4, F_8, FC_5, FC_1, FC_2, FC_6, T_7, C_3, C_z, C_4, T_8, CP_5, CP_1, CP_2, CP_6, P_7, P_3, P_z, P_4, P_8, O_1, O_z, and O_2 electrodes, based on the international 10–10 system at a sampling frequency of 500 Hz with filtering from 0.5 to 120 Hz measured using an amplifier (Neurofax EEG-1100, Nihon Koden Corporation) and active electrodes (actiCap, Brain Products). Linked reference electrodes were placed on both ears of the subjects. In addition to the EEG measurement for the IAT, other types of EEGs were performed with the subjects instructed to blink intentionally and move their eyes upwards; the measurements were used to create templates for calculating the feature value for clustering.

Fourteen (14) healthy men (aged 21–22 years) participated in the experiments. All participants provided written, informed consent and the IAT study design was approved by the Ethics Committee of Tohoku Gakuin University.

3 Analysis Method

To achieve the purpose of this study, we improved and modified the previous method [7] by changing the criteria for the identification of ICs and creating event-related potentials (ERPs) for IAT data in which IC-related artifacts were rejected, instead of artificial EEG noise.

1. Templates were created for eye-blinking and eye movement; an epoch was extracted referring to the vertex of the wave related to the eye-blinks and eye movements, and 20 epochs were averaged for each EEG channel and for each subject.
2. Principal component analysis (PCA) was performed for dimension reduction to prevent overlearning in the ICA of the raw IAT data.
3. ICA for preprocessing data was performed after the PCA procedure had been completed.
4. For each IC, three types of feature values, kurtosis, as well as the cross-correlation coefficient, were evaluated between each IC and the template of the eye-blink and eye movement.
5. K-means clustering was performed by changing the initial centroid ten times in each iteration. The number of clusters K was determined by identifying the largest value in the silhouette coefficients using silhouette analysis.
6. The ICs in the cluster were identified, including the first IC and another single IC comprising another cluster as an artifact, as recommended in a previous study [7]. In subjects 13 and 14 in blocks 4 and 7, almost all the ICs in a cluster, including the first IC, were selected. The exception was the second IC. In this case, we added the criterion to select only the second IC.

7. Reconstruction of the EEG was performed using inverse ICA and then inverse PCA, and it was filtered from 1 to 40 Hz after the rejection of the identified ICs in procedure 6.
8. The epoch for data collection was placed between a prestimulus period of 500 ms and a poststimulus period of 1000 ms at the time of presenting the stimulus of an image or a character.
9. The epochs were averaged 120 times at the maximum to derive the ERPs; an epoch was removed if artifacts over 50 μV in amplitude occurred in the epoch.

4 Results and Discussion

Table 1 summarizes the combinations of ICs identified as artifacts, the number of clusters K, and the channels after PCA, for each subject and block. In some blocks of some subjects, only the first IC was identified; for others, multiple ICs were identified, including the first IC identified as an artifact for subjects 1–12. However, subjects 13 and 14 showed only the second IC in blocks 4 and 7, but not the first IC. This occurred as a result of applying the criteria described in section 3, procedure 6. The result suggests that the artifacts in an artificial EEG mainly project the first IC; however, the artifacts in an EEG during a cognitive task, such as an IAT, can be represented by one or more ICs in addition to the first IC. The results for subject 14 in block 3 suggest that the ICs were inappropriately identified using the criteria, as almost all the ICs (18 of the 22 ICs) were identified. This result is considered to have occurred because the EEG data in subject 14 were contaminated by few eye-blinks and eye movements.

The average number of 120 epochs is displayed in Fig. 1, for the case when ERPs were calculated by first rejecting the identified ICs and then the epochs with artifacts exceeding 50 μV. The average values represent an improvement compared with those obtained from the conventional method [11], which only rejected epochs that included artifacts exceeding 50 μV but did not apply the clustering method based on ICA, in all subjects (including subject 13) except subject 14.

The ERPs at the frontal and occipital areas were analyzed to compare the performances of the proposed and conventional methods. The ERPs obtained from subject 8 at the frontal (F_{p1}) and occipital areas (O_z) showed a large difference in the averaging numbers between the proposed (120 times) and conventional methods (25 times), as shown in Fig. 1. The proposed method displayed an improved signal-to-noise ratio compared with the conventional method in Fig. 2. A large difference was observed in the ERPs (after approximately 300 ms) obtained from the two methods. The ERP peak components were clearly confirmed at the negative-150 ms, positive-200 ms, negative-300 ms, and positive-400 ms components at the F_{p1} electrode in the proposed method.

The ERP components were consistent with those obtained in another IAT–ERP study [12]. This trend in the ERPs was observed in all subjects except subjects 11, 12, and 14.

Table 1 Identified components, number of clusters K, and number of channels after PCA

	Identified component (Number of clusters K, Number of channels after PCA)			
	Block 3	Block 4	Block 6	Block 7
Sub.1	1 ($K = 2$, PCA:17ch)	1 ($K = 2$, PCA:20ch)	1, 12, 16 ($K = 4$, PCA:21ch)	1 ($K = 2$, PCA:20ch)
Sub.2	1, 3 (3, 23)	1, 8 (3, 24)	1 (2, 24)	1, 9, 20 (4, 24)
Sub.3	1, 8 (2, 17)	1, 8, 15 (5, 17)	1 (2, 14)	1 (2, 15)
Sub.4	1, 7, 21 (4, 22)	1, 9 (3, 19)	1, 9, 16, 19 (5, 19)	1, 14 (4, 18)
Sub.5	1 (2, 21)	1, 2 (3, 22)	1 (2, 20)	1 (2, 21)
Sub.6	1, 6, 9, 12, 16, 18, 20, 22 (2, 22)	1, 11, 13, 14, 15, 17, 18, 19 (3, 21)	1, 8, 11, 13, 14, 16, 17 (5, 19)	1, 3, 4, 6, 7, 10, 11, 12, 13, 15, 16, 18 (5, 19)
Sub.7	1, 2 (2, 18)	1, 2 (3, 17)	1 (2, 15)	1 (2, 14)
Sub.8	1, 2 (3, 15)	1, 3 (3, 15)	1, 3, 9 (4, 11)	1, 2 (3, 16)
Sub.9	1 (2, 19)	1 (2, 18)	1, 4 (3, 17)	1 (2, 18)
Sub.10	1, 2 (3, 16)	1, 2, 3 (4, 12)	1, 2 (3, 11)	1, 2 (3, 11)
Sub.11	1 (2, 17)	1 (2, 18)	1 (2, 15)	1 (2, 16)
Sub.12	1 (2, 14)	1 (3, 15)	1 (2, 17)	1 (2, 16)
Sub.13	1 (2, 19)	2 (2, 19)	1 (2, 19)	2 (2, 19)
Sub.14	Not identified (5, 22)	2 (2, 22)	1, 2, 7, 11, 21, 22, 23 (4, 23)	3 (2, 23)

The ERPs for subject 11 are displayed in Fig. 3. It is clear that the average value is insufficient for subject 11. This value is respectively 35 times and twice the average values in the proposed and conventional methods, as shown in Fig. 1. The signal-to-noise ratio was insufficient, owing to the low averaging numbers, even in the proposed method. The same trend was also observed for subject 12. A possible reason for this result is that noise was generated from other sources, apart from eye-blinking and eye movement, such as body movements and stress generated while conducting the experiment.

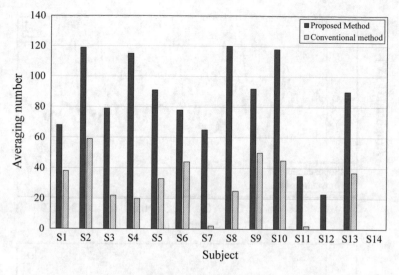

Fig. 1 Averaging number of 120 epochs at the maximum of ERPs

5 Conclusion

The results from this study demonstrate the effectiveness of rejecting artifacts based on suitable ICs, which can be used for visually cognitive tasks, such as IATs. However, the criteria for identifying the ICs in the cognitive task must be expanded to other ICs, where the second IC may be one of the candidates for the artifacts instead of the first IC. The results suggest that the proposed method prevents the incorrect interpretation of ERPs in cognitive tasks. These results can also be employed to select suitable ICs for both real and artificial EEG data.

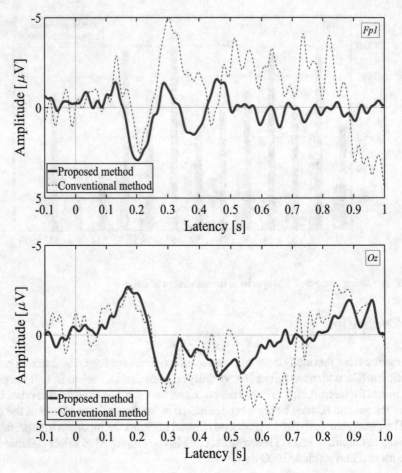

Fig. 2 EPP obtained from subject 8 using the proposed (red curve) and conventional methods (dotted blue curve) at the frontal (F_{p1}, upper plot) and occipital channels (O_z, lower plot)

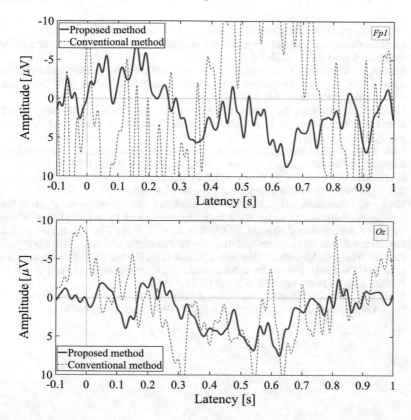

Fig. 3 EPP obtained from subject 11 using the proposed (red curve) and conventional methods (dotted blue curve) at the frontal (F_{p1}, upper plot) and occipital channels (O_z, lower plot)

Acknowledgements The authors declare that they have no conflict of interest. This study was supported in part by a research grant from JSPS KAKENHI Grant Number JP17K00385.

References

1. Delorme, A., Makeig, S.: EEGLAB: an open-source toolbox for analysis of single-trial EEG dynamics including independent component analysis. J. Neurosci. Methods **134**(1), 9–21 (2004)
2. Delorme, A., Palmer, J., Onton, J., Oostenveld, R., Makeig, S.: Independent EEG sources are dipolar. PLoS ONE **7**(2), e30135 (2012)
3. Jung, T.P., Makeig, S., Westerfield, M., Townsend, J., Courchesne, E., Sejnowski, T.J.: Removal of eye activity artifacts from visual event-related potentials in normal and clinical subjects. Clin. Neurophysiol. **111**(10), 1745–1758 (2000)
4. Lindsen, J.P., Bhattacharya, J.: Correction of blink artifacts using independent component analysis and empirical mode decomposition. Psychophysiology **47**(5), 955–960 (2010)

5. Plöchl, M., Ossandón, J.P., König, P.: Combining EEG and eye tracking: identification, characterization, and correction of eye movement artifacts in electroencephalographic data. Front in Human Neurosci **6**, 1–23 (2012)
6. Kanoga, S., Mitsukura, Y.: Proposing an eye blink artifact rejection technique from single-channel EEG signal using positive semi-definite tensor factorization. IEEJ Trans. Electr. Electron. Eng. **135**(7), 848–855 (2015) ((in Japanese))
7. Hiratsuka, S., Hayasaka, D., Kato, K., Kadokura, H.: Identification method of independent components related to artifacts in electroencephalograms. IEEE J Trans. Electr. Electr. Eng. **14**(12), 1836–1841 (2019)
8. Greenwald, A.G., McGhee, D.E., Schwartz, J.L.K.: Measuring individual differences in implicit cognition: the Implicit Association Test. J. Pers. Soc. Psychol. **74**(6), 1464–1480 (1998)
9. Greenwald, A.G., Nosek, B.A., Banaji, M.R.: Understanding and using the implicit association test: 1. An improved scoring algorithm. J. Pers. Soc. Psychol. 85(2), 197–216
10. Kato, K., Kadokura, H., Kuroki, T., Ishikawa, A.: Event-related synchronization/desynchronization in neural oscillatory changes caused by implicit biases of spatial frequency in electroencephalogram. In: Lhotska, L., et al. (eds.) World Congress on Medical Physics and Biomedical engineering 2018, IFMBE Proceedings 68/2, pp. 175–178 (2019)
11. Kato, K., Miura, O., Shikoda, A., Kuroki, T., Ishikawa, A., Kobayashi, T.: Event-related potential affected by spatial frequencies of background visual pattern during a cognitive task. IEEE J. Trans. Electr. Electr. Eng. **8**(5), 483–488 (2013)
12. Healy, G.F., Boran, L., Smeaton, A.F.: Neural patterns of the implicit association test. Front. Human Neurosci. **9**, 605 (2015). https://doi.org/10.3389/fnhum.2015.00605

HydroGEV: Extracellular Vesicle-Laden Hydrogel for Wound Healing Applications

Qingyu Lei, Thanh Huyen Phan, Phuong Le Thi, Christine Poon, Taisa Nogueira Pansani, Irina Kabakowa, Bill Kalionis, Ki Dong Park, and Wojciech Chrzanowski

1 Introduction

Chronic wounds are a substantial social and economic burden for the healthcare system. The global wound management market was estimated at $19.8 Billion USD in 2019, and is projected to reach $24.8 Billion USD in 2024 [1]. The substantial economic burden of wounds arises from the long and costly healing process, and the frequent requirement of complex medical interventions and wound healing aids to promote the healing process.

Q. Lei · T. H. Phan · T. N. Pansani · W. Chrzanowski (✉)
Faculty of Medicine and Health, Sydney School of Pharmacy, Sydney Nano Institute, The University of Sydney, Camperdown 2006, Australia
e-mail: wchrzanowski@sydney.edu.au

C. Poon · I. Kabakowa
Faculty of Science, School of Mathematical and Physical Sciences, University of Technology Sydney, Ultimo 2007, Australia

C. Poon
Faculty of Engineering and IT, School of Biomedical Engineering, University of Technology Sydney, Ultimo 2007, Australia

B. Kalionis
Department of Maternal-Fetal Medicine Pregnancy Research Centre, University of Melbourne, Melbourne, Australia

Department of Obstetrics and Gynaecology, Royal Women's Hospital, Parkville, VIC 3052, Australia

P. Le Thi · K. D. Park
Department of Molecular Science and Technology, Ajou University, Suwon 16499, Korea

T. N. Pansani
Department of Dental Materials and Prosthodontics, Araraquara School of Dentistry, UNESP–University Estadual Paulista, Araraquara, Centro 14801-903, Brazil

© Springer Nature Switzerland AG 2021
C. T. Lim et al. (eds.), *17th International Conference on Biomedical Engineering*, IFMBE Proceedings 79, https://doi.org/10.1007/978-3-030-62045-5_8

Dressings are a common standard of care approach for chronic wounds, which come in different forms ranging from: (i) traditional wound dressings (i.e. gauze, lint, plasters, bandages) as primary or secondary dressings to protect the wound from contamination, (ii) modern wound dressings (i.e. hydrogel, hydrocolloid) that control the wound environment and interact with the wound surface, (iii) bioactive wound dressings (i.e. collagen, hyaluronic acid (HA) that deliver active substances to aid wound healing [2–4]. Unfortunately, current 'gold-standard' dressings do not always result in timely, complete and scar-free wound closure [5]. To address the challenge of chronic wound healing challenge, radical new approaches that integrate multiple active biological and microenvironmental factors into the normal healing process are required [3].

Gelatin is a major protein of the extracellular matrix of skin, bones, and connective tissue, and is the most commonly used component of hydrogels for wound management [6, 7]. Gelatin has been shown to be biocompatible, biodegradable and has good adhesive properties. Gelatin is commercially available at low cost, and has low immunogenicity, which makes it the material of choice for tissue repair, including wound healing [8]. However, like most natural polymers, gelatin gels only within a narrow temperature range. Gelatin-based hydrogel melts below body temperature at ~35 °C, which is makes it unsuitable for would healing [9]. To overcome this problem, Yunki et al. utilized horseradish peroxidase (HRP)-catalyzed cross-linking to form gelatin-hydroxyphenylpropionic acid (GHPA) hydrogel. GHPA hydrogel not only exhibits better temperature stability but the crosslinking level allows modulation of its mechanical properties, which is critical to establish a hydrogel that matches tissue properties and provides appropriate mechanical cues to cells. In addition, GHPA hydrogel has the desired tissue adhesiveness due to enzymatic cross-linking between the phenol groups of gelatin and tyrosine residues of the tissue [10, 11]. Together, these properties make GHPA hydrogel a promising candidate for wound healing. However, to further improve the applicability of GHPA hydrogel for wound healing, other biological cues, such as growth factors and microRNAs (miRNAs) can be incorporated, which together with mechanical cues can enhance and expedite functional tissue repair [12].

Exogenous mesenchymal stromal/stem cells (MSCs) have also been established to have positive therapeutic effects in promoting wound healing [13]. MSCs are currently undergoing clinical trials [14]. However, stem cells must be first harvested, expanded and then transplanted to the injury site, which takes substantial amount of time and delays healing. Furthermore, storing (freezing) stem cells may reduce their survival. Notably, the efficacy of MSCs is hampered by the limited transfer of MSCs to the primary disease sites and inefficient tissue engraftment [15, 16]. Since paracrine factors secreted by MSCs, particularly extracellular vesicles (EVs), possess similar therapeutic effect to whole MSCs [17]. EVs are lipid membrane-enclosed nanoparticles released by cells and they contain multiple biomolecules (e.g. proteins, lipids, DNA, mRNA and miRNA) that reflect their cellular origin [18, 19]. EVs are internalized and unpackaged by recipient cells, which in turn induces their specific biological activities [20]. EVs aid wound healing and suppress inflammation due to burn injuries [21, 22]. Importantly, EVs can be easily harvested and stored in large quantities and readily available for medicinal applications. Therefore, MSCs-derived EVs can overcome current limitations with cell-based therapies and bring about new alternative approach for stem cell therapy and may represent an innovative therapeutic strategy for skin injury repair.

In this study, we combined EVs with GHPA hydrogel to create an advanced EV-laden GHPA hydrogel (HydroGEV) dressing for wound healing applications. We first demonstrated the effectiveness of EVs in promoting cell migration. We then incorporated EVs into GHPA hydrogels and evaluated cell attachment and morphology, cell migration, cell viability and proliferation using human dermal fibroblasts.

2 Methods

2.1 Cell Culture

Maternal MSCs from human term placenta were transduced with human telomerase reverse transcriptase to create DMSC23 cell lines according to protocols described in [23]. DMSC23 cells were cultured in Mesencult™ MSC Basal Medium, 10% stimulatory supplement in Basal medium (Human) (STEMCELL Technologies, Canada), 1% GlutaMAX™ (100× supplement, Life Technologies, France/Japan) and 1% PenStrep. Human dermal fibroblasts (HDF) were cultured in medium containing Dulbecco's Modified Eagle's Medium (DMEM medium–high glucose, Sigma-Aldrich, United Kingdom), 10% foetal Bovine Serum (FBS, Bovogen, Australia) and 1% PenStrep. Hanks' Balanced Salt Solution (HBSS(-), Sigma-Aldrich) was used for washing DMSC23. Phosphate buffered saline (PBS, 10× , Sigma-Aldrich) was used for washing HDF. TrypLE™ Express (gibco, Denmark) was used as the dissociation reagent for adherent cells. All cell lines were maintained at 37 °C and 5% CO_2.

2.2 DMSCs-EVs Isolation and Characterization

Decidual mesenchymal stem cells (DMSC23) were cultured with Mesencult™ MSC culture medium. After cells reached 80% confluency, they were washed twice with HBSS(-) buffer (Sigma Aldrich). After rinsing, cells were maintained for 48 h in Mesencult™ MSC Basal Medium containing 0.5% bovine serum albumin (BSA, Sigma-Aldrich, USA) and 1% PenStrep – EV isolation medium. After 48 h, EV-enriched medium was extracted and transferred into an RNase-free centrifuge tube then centrifuged for 5 min at 500× g to remove apoptotic bodies and detached cells. The supernatant was collected and further centrifuged for 10 min at 2000× g and then passed through a 0.45 μm filter (SFCA membrane, Corning®). EVs were then isolated and concentrated by tangential flow filtration (TFF-EASYTM,

Lonza). EV morphology was then assessed using atomic force microscopy (nanoIR, AnasysInstruments, USA). The size, size distribution and concentration of EVs were measured using a nano flow analyzer (NanoFCM, Xiamen, China).

2.3 Analysis of Cell Migration: XCELLigence Real-Time Cell Analysis (RTCA)

HDFs were cultured in serum-free DMEM medium and 1% PenStrep for 24 h. Cell migration was measured by using xCELLigence® RTCA DP instrument (ACEA Biosciences Inc.) and electronically-integrated Boyden chambers (CIM-Plate 16, China), where the apical side of the top chamber is a microporous membrane, while the basal side was coated with gold microelectrode sensors. Addition of a chemoat-tractant to the bottom chamber induces cells to migrate through the membrane and adhere to the then adhere to the microelectrode coatings in the bottom chamber. The presence of cells alter the impedance signal which is detected and measured by the xCELLigence system as a quantitative kinetic measure of cell status, relative number and migration [24]. To initiate the cell migration test, EVs were added to SFM at a concentration of 1×10^4 EVs per cell. 160 μL of the medium was then added to the bottom chamber, and 30 μL SFM was added to each of the top chambers. 160 μL SFM medium (no EVs) was added to the control groups. The CIM-16 plate was then locked and incubated in the RTCA DP instrument at 37 °C and 5% CO_2 for 60 min. A background signal was recorded when the plate reached equilibrium. At equilibrium, 2×10^4 HDF cells in 120 μL SFM were seeded in each top chamber and incubated at room temperature for 30 min to allow the cell to settle. Finally, the plate was loaded to t measured on the RTCA DP instrument.

2.4 Synthesis of GHPA

GHPA gels were synthesized according to the protocols previously reported by collaborators in [11]. Briefly, gelatin was dissolved in deionized water at 40 °C. Hydroxyphenyl propionic acid (HPA) was activated by 1-ethyl-3-(3-dimethyl aminopropyl)-carbodiimide (EDC) and N-hydroxy-succinimide (NHS) in a co-solvent of H2O and dimethylformamide (DMF) at a volume ratio of 3:2. The solution was mixed with dissolved gelatin. The GHPA solution was stirred for 48 h at 40 °C then collected into a dialysis bag (MWCO. 3.5 kDa) to dialyze three days against deionized water. The GHPA polymer was obtained by filtering and lyophilizing the resulting solution.

Table 1 GHPA hydrogel with different crosslinking densities

Sample name	GHPA conc. (wt.%)	HRP conc. (mg/mL)	H_2O_2 conc. (wt.%)	Stiffnesses
GHPA 1	5.55	0.03	0.075	Low
GHPA 2	5.55	0.03	0.2	Medium
GHPA 3	11.11	0.04	0.3	High

2.5 Formation of GHPA Hydrogel and EV-Laden GHPA Hydrogel–HydroGEVs

GHPA was dissolved in Dulbecco's Phosphate-Buffered Saline (DPBS; Thermo Fisher) at concentrations of 5.55 and 11.11 (wt.%) and made up into two precursor gels for additive curing: (A) one aliquot each with or without EVs was mixed with 0.03–0.04 mg/mL horseradish peroxidase (HRP type VI, 250–330 U/mg solid, Sigma-Aldrich), and (B) one aliquot was mixed with 0.075–0.03% (w/v) hydrogen peroxide (H_2O_2 30 wt. % in H_2O, Sigma-Aldrich) where the concentration of H_2O_2 is directly correlated to crosslinking density and therefore stiffness (Table 1). Gels A and B were then dispensed in a 1:1 ratio. The final concentration of EVs in Hydro-GEVs was 1×10^4 EVs per HDF cell. Longitudinal modulus (frequency, GHz) was measured in a petri dish by Brillouin microscopy to determine relative viscoelastic properties as a proxy for stiffness. Brillouin measurements were taken at 660 nm and $20\times$ confocal objective on a tandem Fabry–Perot interferometer-based system.

2.6 LIVE/DEAD Cell Viability Assay

Cell viability within hydrogel was evaluated by Calcein-AM and propidium iodide (PI) (Live/Dead Assay, Sigma Aldrich) staining. 40 µL GHPA hydrogels with or without EVs were evenly dispensed into a 96-well plate then incubated for 30 min at 37 °C. The crosslinked hydrogels were then rinsed twice with PBS. After rinsing, 1 $\times 10^4$ HDFs mixed in 100 µL of DMEM culture medium were seeded on the surface of the hydrogel. After incubating for 24 h, the gel cultures were rinsed twice with PBS and incubated in 100 µL of the assay solution containing 2% Calcein-AM and 1% PI in PBS in each well maintained at 37 °C for 30 min. After staining, the gel cultures were imaged using a fluorescence microscope (Nikon Eclipse Ti-II inverted microscope).

3 Results and Discussion

Surface morphology analysis using AFM showed that EVs were spherical and their size ranged between 50 and 160 nm (Fig. 1a). Size distribution and concentration

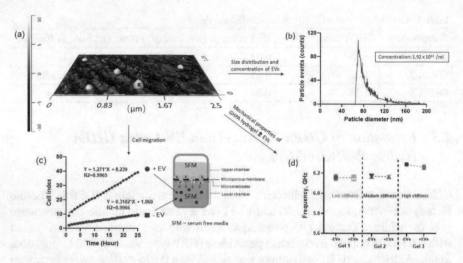

Fig. 1 Physico-chemical characterization of EVs and HydroGEV: **a** size and morphology of individual EVs by AFM; **b** size distribution and concentration of EVs by NanoFCM; **c** cell migration determined by impedance using xCELLigence; **d** mechanical properties assessment by Brillouin frequency shift of GHPA hydrogel with or without EVs

measurements using nano flow analyzer confirmed that size of EVs was between 50 and 200 nm, while the concentration was 1.92×1011 EVs/mL (Fig. 1b).

To demonstrate the effectiveness of EVs in promoting cell migration, we quantified the cell index (CI) using impedance measurements in the xCELLigence cell invasion and migration assay. The CI is a measure of cell translocation through a membrane and allows for real-time monitoring of cell migration. Therefore, by placing EVs in the bottom compartment of the CIM-plate we were able to determine the chemoattractive capacity of EVs. We showed that EVs substantially increased cell migration (Fig. 1c). After 24 h, the CI was 10 without EVs and 40 with EVs, i.e. more cells were attached on the electrodes when EVs presented in the bottom chambers. These impedance measurements suggested that EVs increased the proliferation of HDFs.

However, the efficacy of the EVs can be reduced depending on the microenvironment. In particular, the microstructure and stiffness of extracellular matrix impacts on cell attachment and migration. Therefore, we incorporated EVs into GHPA hydrogels of different stiffnesses that mimic different skin structures. First, we demonstrated that all samples underwent complete gelation within 30 min at 37 °C hydration with DPBS and their macroscopic stiffness was not altered by the presence of the EVs (Fig. 1d).

To determine differences in mechanical properties of the hydrogels we conducted Brillouin microscopy measurements. Brillouin frequency shift directly correlates with the longitudinal modulus and viscoelastic elastic properties (stiffness) of materials, where the higher the frequency shift the higher the longitudinal modulus or stiffness. The Brillouin frequency shifts measured for the hydrogels were 6.156 ±

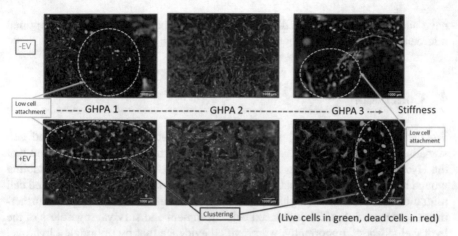

Fig. 2 Cell attachment, distribution and viability by LIVE/DEAD staining assay (scale bar: 1000 μm)

0.030 GHz, 6.171 ± 0.019 GHz and 6.287 ± 0.014 GHz for low, medium and high stiffness gels without EVs, and 6.154 ± 0.029 GHz, 6.166 ± 0.014 GHz and 6.258 ± 0.022 GHz with EVs respectively (Fig. 1d), which correlated with the quantity of hydrogen peroxide and HRP enzyme-mediated crosslinking of the GHPA hydrogels.

The assessment of the cell attachment and their morphology by immunostaining, confirmed high cell survival and proliferation in the HydroGEV regardless of hydrogel stiffness and the presence of EVs (Fig. 2). The cell density was substantially higher in GHPA 2 (medium stiffness). Furthermore, cells on GHPA 2 showed the characteristic elongated fibroblast morphology (Fig. 2). In contrast, the density of cells on GHPA 1 (low stiffness) and 3 (high stiffness) was lower and cell were rounded. Overall, the density of cells increased when EVs were incorporated into all GHPA hydrogels. However, for GHPA 1 and 3, cells were not uniformly dispersed and were presented as clusters (Fig. 2). The results showed that hydrogel stiffness affects cell attachment, specifically, GHPA 2 (medium stiffness) (calculated theoretical stiffness ~7 kPa) showed the highest level of cell attachment. Given that GHPA 2 most effectively supported cell proliferation and uniform cell distribution, it was concluded that GHPA 2 had the most favorable polymer concentration and crosslink density. GHPA 1, on the other hand, had lower crosslinking density (calculated theoretical stiffness ~1 kPa), which appeared to be insufficient for cells to uniformly attach and migrate, which resulted in clustering. While, GHPA 3 had greater polymer concertation (calculated theoretical stiffness ~15 kPa), which was likely to decrease hydrogel porosity. Along with higher stiffness, the microstructure, i.e. porosity, physically limited cell migration through a dense polymer network. However, cell viability, attachment within the hydrogel matrix and the ability of cells to migrate was improved by the incorporation of EVs regardless of the hydrogel stiffness. Nevertheless, a substantial improvement in cell responses was observed when EVs were incorporated into GHPA 2 hydrogel. These results suggest that both mechanical and biological cues have additive or synergistic effects on cells and cumulatively provide an optimal microenvironment for cell growth.

4 Conclusion

We developed a HydroGEV, which combined gelatin-based hydrogel and cell signaling extracellular vesicles. Using a cell culture model, we demonstrated that the HydroGEV has the potential to be an effective dressing material for promoting wound healing. We showed that EVs are strong chemoattractants and increased cell migration four-fold, which highlighted their potential in tissue regeneration. Furthermore, we showed that EVs enhanced cell attachment and survival regardless of the hydrogel stiffness. Importantly, we provided evidence that by optimizing hydrogel stiffness it was possible to promote cell proliferation and achieve the characteristic cell morphology of human dermal fibroblasts, and that the presence of EVs further increased cell migration.

References

1. The Global Wound Care Market Is Projected to Reach USD 24.8 Billion by 2024 from USD 19.8 Billion in 2019, Growing at a CAGR of 4.6%. Plus Company Updates. Plus Media Solutions, October 5, 2019. https://link.gale.com/apps/doc/A601910211/ITOF?u=usyd&sid=ITOF&xid=f4249b2f (Publisher: Plus Media)
2. Schoukens, G.: 5—Bioactive dressings to promote wound healing. In: Rajendran, S. (ed.) Advanced Textiles for Wound Care. Woodhead Publishing, pp. 114–152 (2009)
3. Dhivya, S., Padma, V.V., Santhini, E.: Wound dressings—a review. BioMedicine 5(4), 22–22 (2015)
4. Weller, C., Sussman, G.: Wound dressings. Update 36(4), 318–324 (2006)
5. Madaghiele, M., et al.: Polymeric hydrogels for burn wound care: advanced skin wound dressings and regenerative templates. Burns Trauma 2(4), 153–161 (2014)
6. Frantz, C., Stewart, K.M., Weaver, V.M.: The extracellular matrix at a glance. J. Cell Sci. 123(24), 4195 (2010)
7. Wang, X., Yan, Y., Zhang, R.: Gelatin-based hydrogels for controlled cell assembly. In: Ottenbrite, R.M., Park, K., Okano, T. (eds.) Biomedical Applications of Hydrogels Handbook. Springer New York, New York, NY, pp. 269–284 (2010)
8. Rose, J.B., et al.: Gelatin-based materials in ocular tissue engineering. Materials (Basel, Switzerland) 7(4), 3106–3135 (2014)
9. Lee, Y., et al.: In situ forming gelatin-based tissue adhesives and their phenolic content-driven properties. J. Mater. Chem. B 1(18), 2407–2414 (2013)
10. Jaipan, P., Nguyen, A., Narayan, R.J.: Gelatin-based hydrogels for biomedical applications. MRS Commun. 7(3), 416–426 (2017)
11. Lee, Y., et al.: Enzyme-catalyzed in situ forming gelatin hydrogels as bioactive wound dressings: effects of fibroblast delivery on wound healing efficacy. J. Mater. Chem. B 2(44), 7712–7718 (2014)
12. Wu, Z., et al.: Use of decellularized scaffolds combined with hyaluronic acid and basic fibroblast growth factor for skin tissue engineering. Tissue Eng. Part A 21(1–2), 390–402 (2015)

13. Hu, M.S., et al.: Mesenchymal stromal cells and cutaneous wound healing: a comprehensive review of the background, role, and therapeutic potential. Stem Cells Int. **2018**, 6901983–6901983 (2018)
14. Lu, D., et al.: Comparison of bone marrow mesenchymal stem cells with bone marrow-derived mononuclear cells for treatment of diabetic critical limb ischemia and foot ulcer: a double-blind, randomized, controlled trial. Diabetes Res. Clin. Pract. **92**(1), 26–36 (2011)
15. Fischer, U.M., et al.: Pulmonary passage is a major obstacle for intravenous stem cell delivery: the pulmonary first-pass effect. Stem. Cells. Dev. **18**(5), 683–692 (2009)
16. Gnecchi, M., et al.: Paracrine mechanisms in adult stem cell signaling and therapy. Circ. Res. **103**(11), 1204–1219 (2008)
17. Gomzikova, M.O., James, V., Rizvanov, A.A.: Therapeutic application of mesenchymal stem cells derived extracellular vesicles for immunomodulation **10**(2663) (2019)
18. Kalra, H., Drummen, G.P.C., Mathivanan, S.: Focus on extracellular vesicles: introducing the next small big thing. Int. J. Mol. Sci. **17**(2), 170–170 (2016)
19. Cabral, J., et al.: Extracellular vesicles as modulators of wound healing. Adv. Drug Deliv. Rev. **129**, 394–406 (2018)
20. Zou, W., Liu, G., Zhang, J.: Secretome from bone marrow mesenchymal stem cells: a promising, cell-free therapy for allergic rhinitis. Med. Hypotheses **121**, 124–126 (2018)
21. Zhang, B., et al.: HucMSC-exosome mediated-Wnt4 signaling is required for cutaneous wound healing. Stem Cells **33**(7), 2158–2168 (2015)
22. Li, X., et al.: Exosome derived from human umbilical cord mesenchymal stem cell mediates MiR-181c attenuating burn-induced excessive inflammation. EBioMedicine **8**, 72–82 (2016)
23. Qin, S.Q., et al.: Establishment and characterization of fetal and maternal mesenchymal stem/stromal cell lines from the human term placenta. Placenta **39**, 134–146 (2016)
24. Bird, C., Kirstein, S.: Real-time, label-free monitoring of cellular invasion and migration with the xCELLigence system. Nat. Methods **6** (2009)

Explainable and Actionable Machine Learning Models for Electronic Health Record Data

Ming Lun Ong, Anthony Li, and Mehul Motani

1 Introduction

1.1 Machine Learning in Healthcare

Advances in machine learning show immense promise for healthcare applications. Machine learning methods can accurately identify the relationships between input phenotypic variables and output clinical conditions, proving to be a useful tool for the prediction of clinical incidents [1]. In particular, a suite of deep learning frameworks have been successful in predicting clinical incidents from electronic health record (EHR) data, as deep learning can represent complex, non-linear decision functions [2]. Using de-identified EHR data from multiple healthcare databases, Google Health built a set of deep learning models that outperform traditional healthcare prediction tools by around 5% [3]. To parse clinical notes, state-of-the-art natural-language processing (NLP) methods utilise long short-term memory networks (LSTMs) and attention mechanisms to parse clinical notes [4]. Overall, machine learning shows strong promise for applications in the healthcare domain.

M. L. Ong (✉)
Department of Electrical and Computer Engineering,
National University of Singapore, Singapore 117583, Singapore
e-mail: ongminglun@u.nus.edu

A. Li
Ministry of Health, Sengkang General Hospital, Singapore, Singapore

M. Motani
Department of Electrical and Computer Engineering,
National University of Singapore, Singapore 117583, Singapore

© Springer Nature Switzerland AG 2021
C. T. Lim et al. (eds.), *17th International Conference on Biomedical Engineering*,
IFMBE Proceedings 79, https://doi.org/10.1007/978-3-030-62045-5_9

1.2 Explainable Machine Learning

A key drawback of deep learning models are their black-box nature: while deep learning models yield a high accuracy, a key trade-off is a lack of interpretability [5]. Deep neural networks, such as convolutional neural networks, LSTMs and attention-based models have huge parameter sizes and non-linear connections, improving prediction capabilities [2]. However, these methods are inherently not explainable [6]. This is in contrast to traditional statistical models, which are inherently explainable, due to their ability to attribute prediction outcomes to a set of key features.

Explainable Machine Learning is a research area focused on interpreting and understanding the behaviour of machine learning models, responding to the black-box nature of deep learning models [7]. Explainability has been of particular interest to the healthcare domain, to make decisions in the healthcare space transparent and understandable [8]. A set of explainability methods have been developed to circumvent this problem, and these are critical to deconstruct the behaviour of deep learning models and foster trust amongst clinicians using machine learning. Feature attribution methods are a class of explainability methods, which indicate the significance of each feature on a model's prediction. LIME develops an explanation through linearly approximating the effect of perturbations on the original instance, by optimizing for a loss function, L: $\text{argmin}_g L(f, g, \pi_x)$ [9]. The function optimises over the machine learning model f, the explanation method g, and the proximity measure π_x that defines locality around an instance.

The SHAP method is based on Shapley values from game theory. Explanations are created by calculating the change in prediction outcome when conditioned on a particular feature value being present [10]. The feature importance value $\phi_{i,j,f}$ for an instance i, feature j and machine learning model f is calculated for when a feature value $x_{i,j}$ takes the value k:

$$\phi_{i,j} = E[f(x_{i,j})] - E[f(x_{i,j,f})|x_{i,j} = k]. \tag{1}$$

The output is a set of feature importance values for a specific feature which indicate the significance on a model's prediction.

2 Problem Formulation

While current explainability methods are useful, there are shortcomings regarding their application on machine learning models. Firstly, machine learning models may have significantly different explanations, even if accuracies are comparable. Furthermore, explanation methods are provided for a static set of feature values. Lastly, while explanations are useful, they are of limited utility if they do not inform doctors which features are critical to be acted upon to alleviate a particular condition. If doctors do not trust machine learning explanations, they cannot trust the outcome predictions of the models. Thus, we advocate for a clinically-relevant explainability method which make explanations useful for healthcare. Thus, in this paper, we make the following contributions:

- We develop a three-stage framework. In our first stage, we develop a clinical explanation of how feature importances vary across the spectrum of possible feature values;
- In our second stage, we outline the differences across explanations, and propose a global ranking method which outlines significant explanations on a global level;
- In our last stage, we develop clinical recommendations for actionable features, which best represent the target values that best reduce the incidence of disease.

3   Method

In this section, we provide an outline of our proposed framework to extract insights from the dataset.

3.1   Data

We demonstrate analyses on the prediction of a cardiac disease prediction problem, based on MIMIC-III data [11]. MIMIC-III is an open-source clinical database of deidentified health data, containing information about patients' vital signs, laboratory measurements and lab test information. For the specific cardiac disease problem that we have outlined, our dataset consists of 3960 patient instances and 49 features. For the machine learning methods, missing values in the dataset were imputed using the Multiple Imputation by Chained Equations (MICE) method, and both sets are finally pre-processed using min-max normalization. To split the training and test data, we average test accuracies over 10 iterations of a set of ML models, where the model is trained on 2970 training instances and tested on 990 test instances. The accuracies of each model are outlined in Table 1. Explainability methods will be used on these 4 machine learning models to explain important features.

Table 1 Test accuracy on the LogReg, RF, XGBoost, MLP models

Model	Validation accuracy (%)
Logistic regression (LR)	76.36
Random forest (RF)	78.79
XGBoost (XGB)	80.10
Multilayer perceptron (MLP)	78.08

3.2   Clinical Explainability

We first demonstrate a method to move from individual explanations for a single patient, to a holistic understanding of how feature importances will change when the original feature values are being varied. Firstly, Patient A is selected from the dataset, with the feature values shown in Table 2. The explanations for a single patient are displayed in a plot from the SHAP package, in Fig. 1.

This force plot is generated through the SHAP method. The bar plot highlights the features which are most significant to the prediction. Features which are highlighted in red have a positive contribution to the prediction of heart disease, whereas those highlighted in blue have a negative contribution. Within the context of this explanation, the current SHAP explanation shows that RDW is most essential in predicting heart disease in this patient. While the original explanation provides a general interpretation of key factors in predictive capability, there is more to be desired in real-world clinical applications. A point of interest is the effect on prediction when the feature values change for lower or higher values of haemoglobin.

This premise motivates the development of Clinical Explainability curve. We expand explanations across the set of possible feature values, obtaining an explainability curve that tracks the change of feature importance values. This explanation expansion demonstrates the variance of feature importances over the range of possible feature values, in Fig. 2.

From 2, we demonstrate that a reduction of the RDW value will result in a lower feature importance value. Using this method, the lowest likelihood of disease is achieved when the feature importance value is minimised at RDW = 11.6. As such, this value is set as an optimal **target value** for the particular feature. Additionally, we

Table 2 Feature values for patient A

Feature	Feature value
Age	74
Gender	1
Glucose	95
Creatinine	1.1
Urea nitrogen	24
Potassium	3.8

					base value			higher ⇄ lower output value		
-0.1681	-0.06805	0.03195	0.1319	0.2319	0.3319	0.4319	0.5319	**0.63**9	0.7319	0.8319

22 Extracorporeal circulation auxiliary to open heart surgery = 4.51 | pO2 = 96.77 | Aspirin EC = 325 | Urea Nitrogen = 24 | age = 74 | RDW = 21.4 | Acetaminophen = 650 | Potassium Chlorid

Fig. 1 An explanation of Patient A, using the SHAP package [10] to explain outputs from the random forest model. In this force plot, the predicted probability output by the RF model is 0.63. Features associated with red bars have feature values that positively contribute the output prediction of heart disease, in descending importance: RDW = 21.4, age = 74, urea nitrogen = 24

Fig. 2 A clinical explainability curve for patient A. The patient originally had a RDW value of 21.4, but this curve suggests that a target value of 11.6 is most optimal

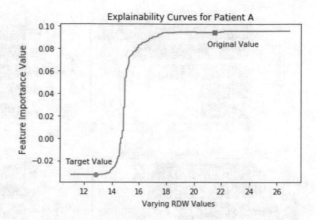

Fig. 3 A clinical explainability curve to explain the SHAP method on the random forest model. This plot of feature importances demonstrates how explanations across the entire dataset can be visualised

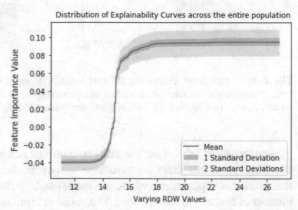

can plot the distribution of explanations for an entire dataset, showing the distribution of feature importance values for a single feature value. For explanations across an entire dataset, we obtain the curve shown in Fig. 3.

3.3 Unifying Across Unstable Explanations

While a set of stable explanations are desired, a key observation made is that explanations are not necessarily consistent. Figure 4 shows the most important features that correspond with a particular model, showing that the most important feature may differ significantly. While age and Red Blood cell Width (RDW) are the two most important features, the order of the third, fourth and fifth important features do not match up. This observation is key in verifying that explanations are not necessarily stable.

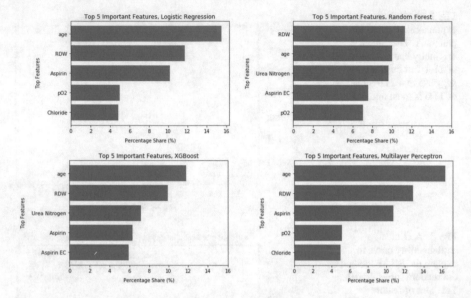

Fig. 4 Most significant features for different machine learning models. To obtain this graph, the feature importance of each feature is taken as a percentage over the sum of all other feature importances. A key observation is the lack of consistency across models

The results in Table 3 outline that explanations across models are not necessarily consistent, and that there are variations across models. We thus outline a method that produces explanations which are model-independent. Firstly, we take in n, the number of features to be explained, as a variable input parameter, where the objective is to find the top n features. The Standard Transferable Vote (STV) is a voting algorithm to achieve proportional representation [12]. Using the STV method, the top n features are selected and the unity of these features present a method to characterise the family of explanations. The output of this segment is a **ranking order** of the most important n features.

3.4 Clinical Actionability

At the end of Sect. 3.2, an optimal value that minimises the likelihood of disease prediction is returned in the target value. In this section, we utilise these target values as sets of actions. An additional constraint is that certain features cannot be acted upon: demographic features, such as age and gender are often important explanations for a clinical condition, but these cannot be modified. Alternatively, there may be a high cost associated with modifying particular features from the medical context, and these features are as such deemed to be not actionable.

Table 3 For patient A, the ranking of most important features SHAP explanations on the LogReg, RF, XGBoost, MLP models, and the aggregated order. The number of explanations characterising this is $n = 8$

Model	LogReg	RF	XGBoost	MLP	Aggregation
1	RDW	RDW	RDW	RDW	RDW
2	Age	Age	Age	Aspirin	Age
3	pO_2	Aspirin EC	Extracorporeal	Admit	Aspirin
4	Aspirin	pO_2	Urea N_2	Age	Aspirin EC
5	Hematocrit	Urea N_2	Aspirin EC	White blood cells	Extracorporeal
6	Hemoglobin	Extracorporeal	Aspirin	Hematocrit	pO_2
7	Nitroglycerin	Calcium Glu.	Hematocrit	Aspirin EC	Nitroglycerin
8	MCHC	Aspirin	Calcium Glu.	Platelet	Dextrose

Table 4 Actions taken on each feature for patient A, according to the most important features on a patient-level. The original feature value is displayed, along with the target value

Rank	Feature	Actionable?	Original	Target	Pred.
1	RDW	No	21.4	12.8	0.630
2	Age	Yes	74	29	0.508
3	Aspirin	Yes	284.35	2.0	0.508
4	Aspirin EC	Yes	325	−29.5	0.478
5	Extracorporeal	Yes	4.51	2.245	0.391
6	pO_2	Yes	96.7	−410.79	0.340
7	Nitroglycerin	Yes	92.8	0.4	0.263
8	Dextrose	Yes	176.38	188.07	0.231

From the ranking order in the previous section, we modify feature values according to the ranking order, and observe the change in output prediction.

For a patient, we outline the change in predicted model output as successive actions are enacted to reduce the patient's likelihood of disease. Table 4 outlines the actions that are taken on a certain patient, and the change in model outcomes.

The analysis shows that the outcome of these patients improve as successive actions are being taken to mitigate a positive disease prediction. When the most significant feature is acted upon, the outcome shifts from 0.613 to 0.231, showing that the change in outcome when this set of actions are followed can reliably reduce the incidence of disease. If the feature values for these important features are successively reduced, the likelihood of positive prediction for these features can be mitigated.

Table 5 Actions taken across the entire population, according to the global ranking. Across the dataset, the most important features are changed to the respective target values. The effect on the average prediction across the dataset is calculated

Rank	Feature	Actionable?	Target	Pred.
1	Gender	No	0	0.331
2	Age	No	35.42	0.331
3	RDW	Yes	12.03	0.295
4	Aspirin EC	Yes	81.58	0.262
5	Hematocrit	Yes	288.72	0.256
6	Urea Nitrogen	Yes	9.66	0.223
7	MCH	Yes	28.56	0.263
8	Aspirin	Yes	105.41	0.231

Additionally, actions can be taken according to the global ranking in Table 5. The important features globally are selected, and the average of target values across all instances' explainability curves are being taken. Across the dataset, all values for the features are being set to the average target value, and the average prediction output of the model is being calculated. Through this analysis, the prediction output can also be reduced.

4   Conclusion

Explainable machine learning methods are a promising method to solve the black-box problem of machine learning and deep learning models. This family of methods can enable trust in practitioners using machine learning to get insights as predictions can be related to a physical variable.

More work on explainable machine learning needs to be done to expand explanations, and address the lack of actionable insight. In our method, we develop a framework that replicates actionability from a clinician. Through the use of clinically-relevant explanations and actionable recommendations, we provide an expansion of explanations which tracks the change in feature importances when feature values are varied, and develop an action plan to help mitigate a patient's disease order.

Finally, additional work can be done to connect machine learning insights against clinical guidelines, so that we can verify if the outcomes will match the clinical outlines suggested by professionals in a medical setting.

References

1. Razavian, N., Marcus, J., Sontag, D.: Multi-task Prediction of Disease Onsets from Longitudinal Laboratory Tests. http://proceedings.mlr.press/v56/Razavian16.html
2. Tu, J.: Advantages and disadvantages of using artificial neural networks versus logistic regression for predicting medical outcomes. J. Clin. Epidemiol. **49**, 1225–1231 (1996)
3. Rajkomar, A., Oren, E., Chen, K., Dai, A., Hajaj, N., Hardt, M., Liu, P., Liu, X., Marcus, J., Sun, M., Sundberg, P., Yee, H., Zhang, K., Zhang, Y., Flores, G., Duggan, G., Irvine, J., Le, Q., Litsch, K., Mossin, A., Tansuwan, J., Wang, D., Wexler, J., Wilson, J., Ludwig, D., Volchenboum, S., Chou, K., Pearson, M., Madabushi, S., Shah, N., Butte, A., Howell, M., Cui, C., Corrado, G., Dean, J.: Scalable and accurate deep learning with electronic health records. npj Dig. Med. **1** (2018)
4. Lipton, Z., Kale, D., Elkan, C., Wetzel, R.: Learning to Diagnose with LSTM Recurrent Neural Networks. https://arxiv.org/abs/1511.03677
5. Ching, T., Himmelstein, D., Beaulieu-Jones, B., Kalinin, A., Do, B., Way, G., Ferrero, E., Agapow, P., Zietz, M., Hoffman, M., Xie, W., Rosen, G., Lengerich, B., Israeli, J., Lanchantin, J., Woloszynek, S., Carpenter, A., Shrikumar, A., Xu, J., Cofer, E., Lavender, C., Turaga, S., Alexandari, A., Lu, Z., Harris, D., DeCaprio, D., Qi, Y., Kundaje, A., Peng, Y., Wiley, L., Segler, M., Boca, S., Swamidass, S., Huang, A., Gitter, A., Greene, C.: Opportunities and obstacles for deep learning in biology and medicine. J. R. Soc. Interface **15**, 20170387 (2018)
6. Jain, S., Wallace, B.: Attention is not Explanation. https://arxiv.org/abs/1902.10186
7. Adadi, A., Berrada, M.: Peeking inside the black-box: a survey on explainable artificial intelligence (XAI). IEEE Access **6**, 52138–52160 (2018)
8. Holzinger, A., Biemann, C., Pattichis, C., Kell, D.: What Do We Need to Build Explainable AI Systems for the Medical Domain? https://arxiv.org/abs/1712.09923
9. Ribeiro, M., Singh, S., Guestrin, C.: Why Should I Trust You? Explaining the Predictions of Any Classifier. https://arxiv.org/abs/1602.04938
10. Lundberg, S., Lee, S.: A Unified Approach to Interpreting Model Predictions. https://arxiv.org/abs/1705.07874
11. Johnson, A., Pollard, T., Shen, L., Lehman, L., Feng, M., Ghassemi, M., Moody, B., Szolovits, P., Anthony Celi, L., Mark, R.: MIMIC-III, a freely accessible critical care database
12. Bartholdi, J., Orlin, J.: Single transferable vote resists strategic voting. Soc. Choice Welfare **8**, 341–354 (1991)

The Impact of Static Distraction for Disc Regeneration in a Rabbit Model—A Longitudinal MRI Study

Wing Moon Raymond Lam⊙, XiaFei Ren, Kim Cheng Tan⊙,
Kishore Kumar Bhakoo, Ramruttun Amit Kumarsing, Ling Liu,
Wen Hai Zhuo, Hee Kit Wong, and Hwee Weng Dennis Hey⊙

1 Introduction

Low back pain resulting from disc degeneration is a disabling condition that imposes a huge socioeconomic burden. Although spinal fusion surgery can provide symptom relief in refractory cases, its benefits come at the expense of accelerating adjacent spinal level degeneration. Intervertebral disc (IVD) regeneration strategies remain ideal and promises the best outcome.

The intervertebral disc is the largest avascular structure in the body and relies on the microvasculature in the adjacent vertebral endplates and peripheral annulus fibrosus to receive nutrients and expel waste products. Transport in the disc is governed by a combination of diffusive and convective transport. From our previous bioreactor studies, distraction can lead to disc regeneration and increase vascularity.

Various studies has been performed in bioreactor [1] or rabbit model to study the effects of static [2] and dynamic distraction [3] on disc regeneration. Unfortunately, the previous study is a single time point study that fail to provide information of long term treatment efficacy on disc health. In addition, the MRI scan is performed ex vivo that cannot mimic the physiological condition [4].

In this longitudinal pilot study, we will develop a MRI-compatible rabbit IVD distraction. This will allow us to evaluate the long-term effects vertebral distraction have on disc health in vivo. The aim of this study is to develop an MRI-compatible,

W. M. R. Lam · R. A. Kumarsing · L. Liu · W. H. Zhuo · H. K. Wong · H. W. D. Hey (✉)
Orthopaedic Surgery, National University of Singapore, Midview City, Singapore
e-mail: doshhwd@nus.edu.sg

X. Ren · K. C. Tan
School of Engineering, Temasek Polytechnic, Tampines, Singapore

K. K. Bhakoo
Translational Imaging (DTI): Singapore Bioimaging Consortium , Agency for Science, Technology and Research (A*STAR), Helios, Singapore

© Springer Nature Switzerland AG 2021
C. T. Lim et al. (eds.), *17th International Conference on Biomedical Engineering*,
IFMBE Proceedings 79, https://doi.org/10.1007/978-3-030-62045-5_10

rabbit IVD distraction model to enable longitudinal monitoring of nutrient supply and disc regeneration.

2   Material and Methods

2.1   Distractor Design

To ensure MRI compatibility for longitudinal study of disc health, our distractor device was modified from the previous study to replace all stainless steel component with MRI compatible plastic and zirconia. Zirconia is chosen due to high MRI compatibility and strength [5]. The modified distractor is composed of (1) 2 Zirconia k-wires to be inserted into adjacent vertebral bodies, (2) 4 Zirconia connecting rods for attaching the K-wires to an external frame, (3) a PEEK frame, and (4) a urethane washer acting as a spring (Fig. 1). The force constant of spring was evaluated by Instron tester.

CT and MRI scans were performed on rabbit carcass model containing the assembled device to ascertain interference. CT scan were performed with Somatom Definition (Siemen, Germany) at 100 kV and 250 mAS. MRI scans was performed with 3 T Magnetom Skyra (Siemen, Germany).

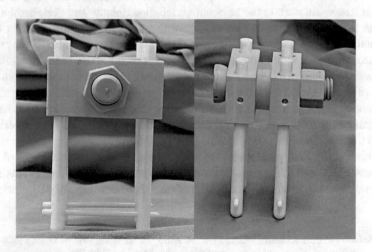

Fig. 1 Picture illustration of the distractor device

2.2 Animal Study

Two male New Zealand white rabbits weighting average of 3.6 kg were used in the current study. All surgical and experimental procedures were conducted under a protocol approved by Institutional Animal Care and Usage Committee, National University of Singapore (R17-1548).

Rabbit IVD stabbing was modified from DW Kim et al. method [6]. Under general anaesthesia (2% isoflurane), all rabbits were stabbed at the L4/5 IVD using a bone marrow aspirate needle with k-wire guidance under C-arm guidance. A baseline MRI scan was performed at 6 weeks to assess for IVD degeneration. After confirmation of disc degeneration, rabbits were divided into two groups—(1) Control (no treatment), and (2) Distraction.

After confirm the degree of disc degeneration by 6 weeks post-stabbing MRI. Under general anaesthesia and strict aseptic technique, distractor device was implanted via a dorsal approach using a 4-5 cm midline incision. After blunt muscle dissection to expose the transverse processes, a 2 mm K-wire was used to drill through the L4 and L5 vertebra bodies under C-arm guidance (Fig. 2). The K-wires were subsequently replaced with zirconia wires (2 mm diameter). Additional skin port incisions were then made to allow passage of connecting rods to the Zirconia wires. Soft tissue and skin was closed layer by layer. Finally, the external PEEK frame and urethane washer were assembled to generate a 120 N tensile force. Rabbit were returned the home cage post-Operation, buprenorphine, carprofen were injected for 1 week to provide necessary pain relief. Enrofloxacin was given as prophylaxis antibiotic to reduce chance of skin port infection.

Fig. 2 C-arm image of K-wires inserted through vertebra bodies

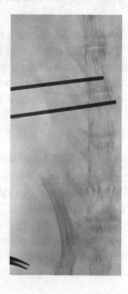

2.3 Disc Health Assessment

The degree of disc degeneration was assessed at 7, 11 and 15 weeks post-distractor device implantation and no treatment group. T2-weighted STIR MRI was used to determine the degree of disc hydration (TR 5000, TE 114, FOV 150 × 150, Flip angle 150 degree, average = 2, slide thickness 1.7 mm, 6 slices). The T2-weighted image slide, one per subject, were used to evaluate NP brightness which is a function of the level of hydration and therefore of degeneration in the disc [7]. Deuterium standard was prepared according Victor Leung et al. publication to determine the variation among each MRI scan [8].

Nutrient flow was determined by T1-weighted pre- and post-contrast MRI. Rabbit were injected with 0.3 ml/kg of T1 contrast Omniscan (GE Healthcare, USA). Time between pre and post contrast is 10 min. A series of TR/TE was used for measurement (TR 250-3000 ms with increment of 500 ms, TE 9.8 ms, FOV 150 × 150 mm, Flip angle 180 degree, average = 1, slide thickness 1.7 mm, 9 slices).

The T1 signal intensity of Pre contrast and post contrast images at different TR and TE were analysed by Syngo.via software. The disc was selected by Volume of interest (VOI) tools. The difference between Pre and Post-Contrast T1 signal was compared.

3 Results

3.1 Implantation Interference

From Fig. 3a, minimal MRI interference was observed following implantation of the 2 mm Zirconia K-wire into the rabbit carcass. With dual X-ray artefact suppression method, wire loosening can also be accurately diagnosed despite the artefact of high density of Zirconia K-wire. Clear black line was identified in loosen K-wire. (Fig. 3b, c).

3.2 Static Distraction Improve Disc Health

MRI at 6 weeks post IVD stabbing confirmed disc degeneration specific to the level of intervention (Fig. 4a, b). The treatment group showed partial restoration of disc hydration on the T2-weighted STIR MRI after 11 weeks of distraction treatment while control group disc hydration level further deteriorate in 17 and 21 weeks post stabbing MRI (Fig. 4c, d). Control group disc become completely dehydrated (dark in T2 MRI image) on 17 weeks post-stabbing.

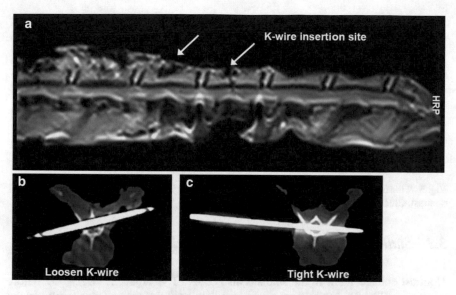

Fig. 3 **a** MRI with implanted Zirconia K-wire; **b** CT image of vertebra with loosen K-wire created by enlarged drilling hole clear gap was shown **c** CT of tight fixed K-wire implant

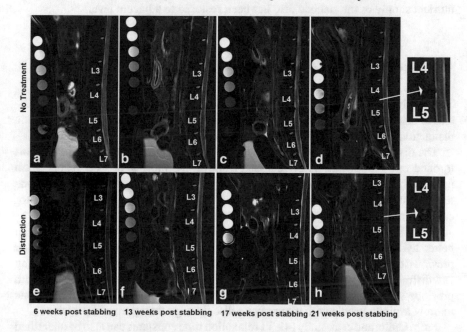

Fig. 4 T2-weighted STIR MRI at baseline (6 weeks post stabbing), 7th, 11th and 15th week post distraction treatment with (below) and no treatment control (above) The sequence of deuterium oxide standard is 0, 20, 40, 60, 80 and 100% (descending order in MRI image)

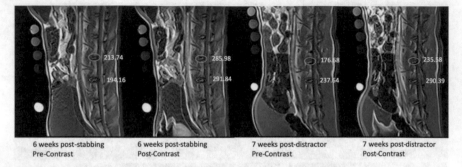

6 weeks post-stabbing   6 weeks post-stabbing   7 weeks post-distractor   7 weeks post-distractor
Pre-Contrast   Post-Contrast   Pre-Contrast   Post-Contrast

Fig. 5 T1-weighted pre- and post-contrast MRI Rabbit was injected with 0.3 ml/kg Omniscan T1 contrast. Green Stabbed disc Red Adjacent disc

3.3 Static Distractor Restored Disc Nutrient Supply

Contrast enhancement in the stabbed disc of + 72 vs adjacent level + 97 before distraction (Fig. 5). After 7 weeks of distraction, both the stabbed disc (+59) and the adjacent level (+53) bear similar contrast enhancement. These results suggest that nutrient supply of the stabbed disc has been restored to adjacent level.

4 Discussion and Conclusion

This in vivo rabbit longitudinal study showed that IVD distraction may be beneficial to disc health if applied over a certain duration. The effect of long-term static distraction to improve disc hydration was level off on 11 weeks post- distraction.

The distraction period of current study is three time longer than previous report (around one month). This new distractor allows monitoring of IVD regeneration process in physiological condition in a longitudinal manner while previous one require euthanizing animal for MRI scan [4]. The adoption of non-magnetic distractor can offer a mean to reduce animal usage and minimize individual difference.

From 11 weeks post-distraction T2 Stir image, rabbit stabbed MRI has shown better hydration than untreated counterpart. The result matched with our group previous bioreactor and rabbit static distractor study that demonstrate unloading and distraction allowed the regeneration of the extracellular matrix. Compare with previous study, our current model was created by stabbing with K-wire and bone marrow aspirate needle with more severe disc damage than pure compression.

From Gullbrand et al. study [4], T1 relaxation time constants can also be quantified for a circular region of interest in the central nucleus pulposus and for elliptical region of interests in the cartilage endplates and anterior and posterior annulus fibrosus. This will provide small molecule (nutrient or waste) uptake and clearance data. With current in vivo model, same rabbit can be used to measure uptake and clearance with physiological condition.

The success of this MRI-compatible model establishes the platform in which future studies can be performed to further evaluate various surgical intervention on disc regeneration.

Acknowledgements The study was supported by NMRC CS-IRG-NIG CNIG may 015.

Reference:s

1. Hee, H.T., Zhang, J., Wong, H.K.: Effects of cyclic dynamic tensile strain on previously compressed inner annulus fibrosus and nucleus pulposus cells of human intervertebral disc-an in vitro study. J. Orthop. Res. **28**(4), 503–509 (2010). https://doi.org/10.1002/jor.20992
2. Hee, H.T., Chuah, Y.J., Tan, B.H., Setiobudi, T., Wong, H.K.: Vascularization and morphological changes of the endplate after axial compression and distraction of the intervertebral disc. Spine (Phila Pa 1976) **36**(7), 505–511 (2011). https://doi.org/10.1097/BRS.0b013e3181d32410
3. Kroeber, M., Unglaub, F., Guehring, T., Nerlich, A., Hadi, T., Lotz, J., Carstens, C.: Effects of controlled dynamic disc distraction on degenerated intervertebral discs: an in vivo study on the rabbit lumbar spine model. Spine (Phila Pa 1976) **30**(2), 181–187 (2005). https://doi.org/10.1097/01.brs.0000150487.17562.b1
4. Gullbrand, S.E., Peterson, J., Mastropolo, R., Roberts, T.T., Lawrence, J.P., Glennon, J.C., DiRisio, D.J., Ledet, E.H.: Low rate loading-induced convection enhances net transport into the intervertebral disc in vivo. Spine J **15**(5), 1028–1033 (2015). https://doi.org/10.1016/j.spinee.2014.12.003
5. Smeets, R., Schollchen, M., Gauer, T., Aarabi, G., Assaf, A.T., Rendenbach, C., Beck-Broichsitter, B., Semmusch, J., Sedlacik, J., Heiland, M., Fiehler, J., Siemonsen, S.: Artefacts in multimodal imaging of titanium, zirconium and binary titanium-zirconium alloy dental implants: an in vitro study. Dentomaxillofac Radiol. **46** (2):20160267 (2017). 10.1259/dmfr.20160267
6. Kim, D.W., Chun, H.J., Lee, S.K.: Percutaneous needle puncture technique to create a rabbit model with traumatic degenerative disk disease. World Neurosurg **84**(2), 438–445 (2015). https://doi.org/10.1016/j.wneu.2015.03.066
7. Videman, T., Gibbons, L.E., Battie, M.C.: Age- and pathology-specific measures of disc degeneration. Spine (Phila Pa 1976) **33**(25), 2781–2788 (2008). https://doi.org/10.1097/BRS.0b013e31817e1d11
8. Leung, V.Y., Hung, S.C., Li, L.C., Wu, E.X., Luk, K.D., Chan, D., Cheung, K.M.: Age-related degeneration of lumbar intervertebral discs in rabbits revealed by deuterium oxide-assisted MRI. Osteoarthritis Cartilage **16**(11), 1312–1318 (2008). https://doi.org/10.1016/j.joca.2008.03.015

Predictive Biomechanical Study on the Human Cervical Spine Under Complex Physiological Loading

S. Dilip Kumar, R. Shruthi, R. Deepak, D. Davidson Jebaseelan, Lenin Babu, and Narayan Yoganandan

1 Introduction

It is well known that the human musculoskeletal structures resist normal day-to-day physiological loads (forces and moments) via deformations (displacements and angulations) while maintaining the integrity of their internal components. For example, the function of the osteoligamentous spinal column is to protect the integrity of the neural structures and preserve the interrelationships between its joints (facet and disc) from the axis to the lower lumbar spine. Degradation of the spine is a normal outcome of the ageing process in the human life—advancing age is associated with increased degenerative qualities and may lead to spinal disorders. Common disorders are spondylosis, radiculopathy (disc herniation), and myelopathy. Anterior cervical spine discectomy (ACDF) is one of the most common operations around the world since the 1950 [1, 2]. The process involves the removal of the diseased

S. Dilip Kumar (✉) · D. Davidson Jebaseelan · L. Babu
School of Mechanical Engineering, Vellore Institute of Technology, Chennai Campus, India
e-mail: dilipdivya98@gmail.com

D. Davidson Jebaseelan
e-mail: davidson.jd@vit.ac.in

L. Babu
e-mail: lenin.babu@vit.ac.in

R. Shruthi
School of Computer Sciences and Engineering, Vellore Institute of Technology, Chennai Campus, India
e-mail: ramachandranshruthi2@gmail.com

R. Deepak · N. Yoganandan
Department of Neurosurgery, Medical College of Wisconsin, Milwaukee, WI, USA
e-mail: drajasekaran@mcw.edu

N. Yoganandan
e-mail: yoga@mcw.edu

© Springer Nature Switzerland AG 2021
C. T. Lim et al. (eds.), *17th International Conference on Biomedical Engineering*,
IFMBE Proceedings 79, https://doi.org/10.1007/978-3-030-62045-5_11

or painful disc and introduction of a graft, and this change in the structure of the spine results in altered biomechanics. On a longitudinal basis, following the operation, doctors/surgeons record the functional status of the spine using angulation as a metric, termed as range of motions (ROM) at the operated (index) and adjacent (caudal and cranial) levels and compare to the pre-operative state, and also monitor the status of the other joints (disc and facets at other levels).

The biomechanical response of the ACDF spine and its comparison to the intact cervical spine can be studied in a laboratory using human cadaver spine experiments and or computational models, i.e., finite element simulations [3]. The reproducibility of the finite element model simulations is particularly suitable for comparing the responses between surgically altered and intact spines. While complex models can incorporate details of the three-dimensional cervical spine anatomy and facilitate more accurate response estimations, they are resource intensive (e.g., nonlinearity and convergence issues). Subject-specific models require intense development.

2 Materials and Methods

2.1 Finite Element Model

Machine learning is an upcoming field wherein surgeons can make use of the history to guide the decision for a new patient. For the machine learning to be effective, it is important to generate a baseline dataset that is validated and then use it for future applications. Training and testing sets are a part of this paradigm. The aims of this study are as follows:

An accurately meshed intact model, a model with ACDF at C5-6 level and another with mild degeneration at C5-6 level were simulated using idealised material properties with corresponding boundary conditions. Data sets generated under physiological loading will be trained using machine learning algorithms.

2.2 Material Properties and Boundary Conditions

Material properties shown in Table 1 were used for the seven segments (C2-T1). Anterior plate was not modelled in order to reduce complexity and hence, this was carried on by altering the material properties of the C5-C6 level natural discs to that of the bone [4]. Moreover, this proves to be as effective as inter-body fusion using graft and plate [5]. Furthermore, the intact model was modified to simulate degeneration at C5-C6 level by assigning elastic values and making it stiffer. The elastic values assigned to the nucleus were twice that of annulus ground [6]. An external load of 2 Nm was applied at the centre of gravity of the C2 vertebrae

Table 1 Elements and material properties used in the study

Component	Element type	Constitutive model	Properties (elastic modulus in Mpa)	Reference
Cortical bone	Hexahedral solid	Linear elastic	E = 16,800, μ = 0.3	[7]
Trabecular bone	Hexahedral solid	Linear elastic	E = 400, μ = 0.3	[8, 9]
Annulus ground	Hexahedral solid	Hyperelastic mooney-rivlin	C10 = 0.56 C01 = 0.14 V = 0.45	[10]
Nucleus pulposus	Hexahedral solid	Hyperelastic mooney-rivlin	C10 = 0.12 C01 = 0.09 V = 0.4999	[10]
Endplate	Quadrilateral shell	Linear elastic	E = 5600, μ = 0.3	[11]
Facet cartilage	Quadrilateral shell	Linear elastic	E = 10, μ = 0.3	[12]
Ligaments	Quadrilateral membrane	Non-linear fabric	Stress–strain curves	[13, 14]

through a kinematic coupling. All degrees of freedom at the bottom endplate of T1 vertebrae were constrained (Fig. 1).

2.3 Python Scripting, Training and Prediction

ABAQUS has a built-in python API that offers many benefits when utilized. ROM, Intradiscal pressure and facet forces were extracted from an Abaqus output database using the Abaqus-python functions and imported into the scripting environment. The creation of multiple job files for several boundary conditions and processing each of them in parallel was made possible with this integrated interface (Fig. 2).

A machine learning algorithm was trained with a dataset of size 150,000 after feature engineering, data pre-processing, and model optimization to create the most accurate predictive model [15]. Random forest regression, support vector regression and multiple linear regression models were performed on the dataset and the results were compared to see the best fit with the least error function.

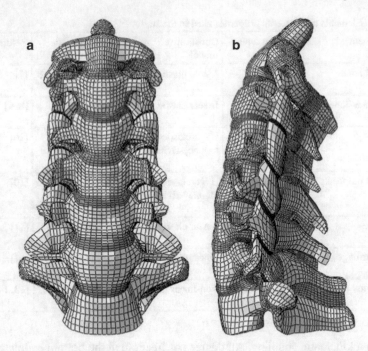

Fig. 1 Intact model front view (**a**) and side view (**b**)

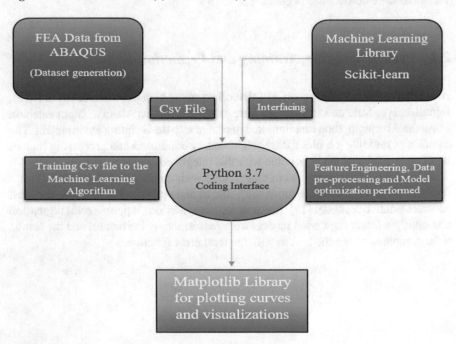

Fig. 2 The predictive algorithm process flowchart

3 Results

3.1 Range of Motion

The model was simulated for sagittal, frontal and transverse plane moment load cases of 2 Nm considering material and geometric non-linearity. The Range of Motion obtained from the intact model simulation was first compared with another finite element model [16] and an experimental study [17]. The most flexible FSU in flexion was the C4-C5 segment exhibiting highest ROM. The lower segments displayed almost equal flexibility. C5-C6 segment was the most flexible segment in extension and C2-C3 segment was stiffer compared to the rest of the model. The ROM for the intact model fell well within literature limits for all the segments for the corresponding load cases as shown in Fig. 3.

Similarly, ACDF and mild degeneration was also simulated for sagittal, frontal and transverse plane moment load cases of 2 Nm. The ROM for the single level ACDF and degenerated model obeyed the trend as shown in Fig. 4.

3.2 Intradiscal Pressure

Change in the intra-disc pressure often leads to uneven load distribution which causes disc degeneration as the nucleus ruptures. FEM results for IDP were compared with the literature values [18] and fell within the corridor. Furthermore, IDP values at the adjacent levels for single-level (C56) ACDF and degeneration models were compared with that of the intact model and plotted in the Fig. 5.

3.3 Facet Force

Results of facet force for flexion, extension, lateral bending and axial rotation for the intact, degenerative and fusion models were within the literature values [19]. Additionally, facet force at the adjacent levels for single-level (C5-C6) ACDF and degenerative models were compared with the intact model and plotted in Fig. 6.

3.4 Predictions from ML Models

A simple multiple linear regression model was first trained for the dataset (intact and single level ACDF and mild degeneration at C5-6 level) where the input features included load (flexion, extension, lateral bending and axial rotation), FSU level, Young's modulus of the natural disc and the type of condition (intact, ACDF or

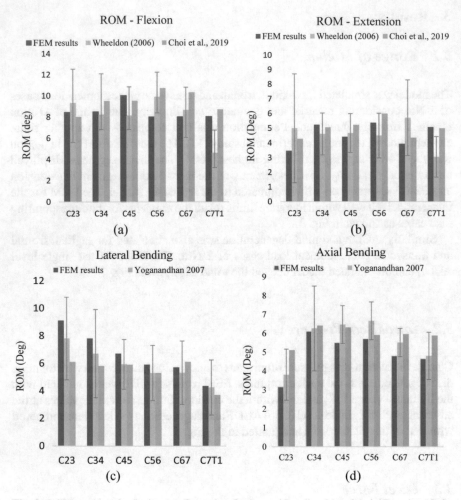

Fig. 3 ROM corridors for flexion (**a**), Extension (**b**), Lateral Bending (**c**) and Axial Rotation (**d**) for intact model

degeneration). A sample test data for a single level ACDF model with mild degeneration occurring at C4-C5 was considered for testing and predicting ROM. Finite element simulation was performed for the same and the corresponding ROM results were compared for accuracy. ROM at C2-C3 with level 1 degeneration at C5-C6 was taken as a test case in order to compare the accuracy of different ML models with the actual simulation results. The comparison plots of different ML models are given in Fig. 7.

The R^2 value for multiple linear regression was found to be 70% with Mean Absolute Error (MAE) and Root Mean Square Error (RMSE) of 1.482 and 1.911 respectively. There was an accuracy drop and error increase observed when a support vector regressor with the RBF kernel was modelled for the same test and training

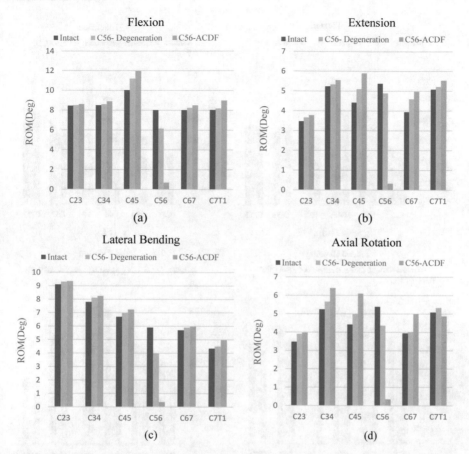

Fig. 4 ROM corridors for flexion (**a**), and extension (**b**), Lateral Bending (**c**) and Axial Rotation (**d**) after ACDF and degeneration at C56

data with R^2, MAE and RMSE to be 18%, 3.932 and 4.802 respectively. However, a random forest regression algorithm [20] outperformed these two models with a steep rise in prediction performance owing to 94% accuracy. MAE was 0.550 and RMSE was 0.837. Random forest regression was also able to perform multi-output regression by default whereas the rest needed separate training for each target variable.

3.5 Visualization and Comparison with FEM

Prediction of complex loading was done using random forest regression by assigning the moment load across different planes as input to our predictive model with degenerative properties across the C4-C5 FSU and ACDF across the C5-C6 FSU for testing the reliability of the predictive model. Different cases which were provided as input

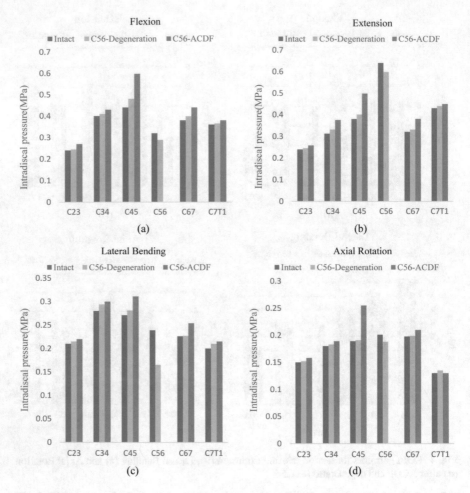

Fig. 5 IDP corridors for flexion (**a**), extension (**b**) lateral bending (**c**), Rotation (**d**) after degeneration at C56

for prediction are tabulated in the table. As the moment load is applied across different planes simultaneously, ROM will be resolved across all the three planes. ROM for each FSU in the corresponding planes are also plotted. Additionally, IDP, Facet force and ROM across the oblique plane are also plotted for the 2Nm moment load. For the initial prediction, combined moment load of 2Nm was applied with the sagittal and frontal plane to simulate complex physiological loading.

Similarly, predictions were done on Axial extension, Axial flexion and lateral extension, FEA simulations were run for the same in order to compare the reliability of the predicted results and were found to be agreeable to that of FEM data with a minimal error and with an accuracy of approximately around 94%. It was also observed that the values of ROM, facet force and pressure were predicted accurately for other conditions of complex loading as well.

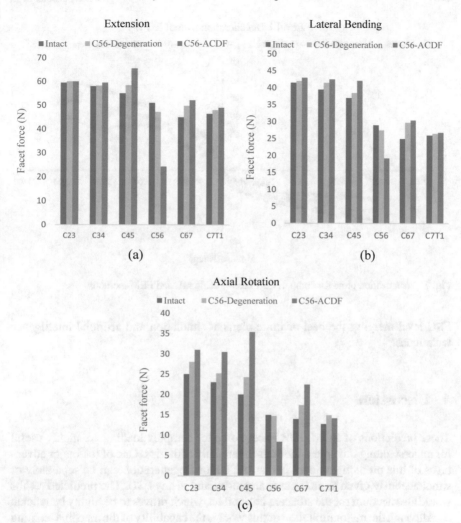

Fig. 6 Facet force corridors for extension (**a**) lateral bending (**b**) Rotation (**c**) after degeneration at C56

A gradual increase in prediction accuracy can be observed from the line graph (Fig. 8) shown above. The results clearly show that the random forest regression was more accurate the test data fed into the model whereas, support vector machines show the least accuracy. Predictive models has led to results with better accuracy, less processing time, power and memory while following literature standards. This work brought out the importance of studying the bio-mechanics of a human cervical spine by understanding the impact of ROM, IDP and facet force of every FSU when either an ACDF or degeneration or a combination of the both has occurred at any

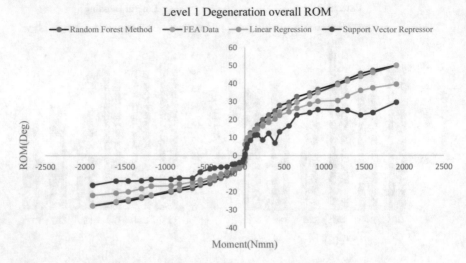

Level 1 Degeneration overall ROM

Fig. 7 Visualization plots for comparison of ROM using ML and FEM methods

FSU level merging the best of finite element simulation and artificial intelligence techniques.

4 Discussion

These predictions of ROM, IDP, Facet Force for complex loading are highly useful for understanding our spinal motions in our daily activities. One of the major advantages of this predictive model is that, ACDF or Degeneration can be separately or simultaneously given as an input at any levels and in any FSU. The predicted results would also account for the adjacent level effect, which proves to be highly beneficial.

Although the major limitation of this work is the capability of the machine learning model to predict results for different genders and ages of the patient's spine, the cervical spine model of only one particular patient is used for generating FEM data. But this can be overcome by training more patient specific data, which broadens the predicting parameters of our machine learning model for extensive results.

The change in ROM after various clinical conditions and subsequent changes in the spine geometry and other intrinsic parameters can be studied using such machine learning algorithms.

5 Conflict of Interest

The authors do not have any known conflicts of interest with regard to this publication.

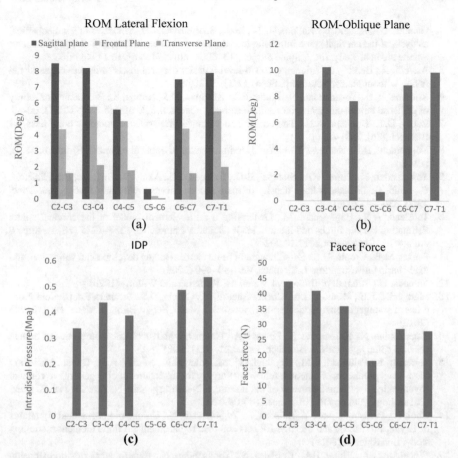

Fig. 8 Projected ROM (**a**), Oblique ROM (**b**) IDP (**c**), Facet force (**d**) for Lateral Flexion

Acknowledgements This study was supported by the Dassault Systemes Foundation, India; The Department of Neurosurgery ,Medical College of Wisconsin, USA.

References

1. Panjabi, M.M., Cholewicki, J., Nibu, K., et al.: Criticial load of the human cervical spine: an in vitro experimental study. Clin. Biomech. **13**, 11–17 (1998)
2. Yoganandan, N., Kumaresan, S., Pintar, F.A.: 'Biomechanics of the cervical spine. Part 2. Cervical spine soft tissue responses and biomechanical modeling.' Clin. Biomech. **16**(1), 1–27 (2001)
3. Merali Zamir, G. et al.: Using a machine learning approach to predict outcome after surgery for degenerative cervical myelopathy. PloS one **14**(4), e0215133 (2019). https://doi.org/10.1371/journal.pone.0215133

4. Gandhi, A., Grosland, N., Kallemeyn, N., Kode, S., Fredericks, D., Smucker, J.: Biomechanical analysis of the cervical spine following disc degeneration, disc fusion, and disc replacement: a finite element study. Int. J. Spine Surgery **13**, 6066. https://doi.org/10.14444/6066

5. Assietti, R., Beretta, F., Arienta, C.: Two-level anterior cervical discectomy and cageassisted fusion without plates. Neurosurg. Focus **12**(1), 3 (2002)

6. Kumaresan, S., Yoganandan, N., Pintar, F.A., Maiman, D.J., Kuppa, S.: Biomechanical study of pediatric human cervical spine: a finite element approach. J. Biomech. Eng. **122**(1), 60–71

7. Reilly, D.T., Burstein, A.H.: The elastic and ultimate properties of compact bone tissue. J Biomech **8**(6), 393–405 (1975)

8. Kopperdahl, D.L., Keaveny, T.M.: Yield strain behavior of trabecular bone. J. Biomech. **31**(7), 601–8

9. Yoganandan, N., Pintar, F.A., Stemper, B.D., Baisden, J.L., Aktay, R., Shender, B.S., Paskoff, G., Laud, P.: Trabecular bone density of male human cervical and lumbar vertebrae. Bone. **39**(2), 336–44 (2006)

10. Toosizadeh, N., Haghpanahi, M.: Generating a finite element model of the cervical spine: Estimating muscle forces and internal loads. Scientia Iranica. **18**, 1237–1245 (2011). https://doi.org/10.1016/j.scient.2011.10.002

11. Panzer, M.B., Cronin, D.S.: C4–C5 segment finite element model development, validation, and load-sharing investigation. J. Biomech. **42**, 480–490 (2009)

12. Yamada, H.: Strength of Biological Materials. Williams and Wilkins (1970)

13. Mattucci, S.F.E., Moulton, J.A., Chandrashekar, N., Cronin, D.S.: Strain rate dependent properties of younger human cervical spine ligaments. J. Mech. Behav. Biomed. Mater. **10**, 216–226 (2012)

14. Yoganandan, N., Kumaresan, S., Pintar, F.A.: Geometric and mechanical properties of human cervical spine ligaments. J. Biomech. Eng. **122**, 623 (2000)

15. Tetreault, L., Palubiski, L.M., Kryshtalskyj, M., Idler, R.K., Martin, A.R., Ganau, M., et al.: Significant predictors of outcome following surgery for the treatment of degenerative cervical Myelopathy: a systematic review of the literature. Neurosurg. Clin. N. Am. **29**(115–27), e35 (2018). https://doi.org/10.1016/j.nec.2017.09.020

16. Choi, H., Purushothaman, Y., Baisden, J., Yoganandan, N.: Unique biomechanical signatures of Bryan, Prodisc C, and Prestige LP cervical disc replacements: a finite element modelling study. Eur. Spine J. (2019)

17. Wheeldon, J.A., Pintar, F.A., Knowles, S., Yoganandan, N.: Experimental flexion/extension data corridors for validation of finite element models of the young, normal cervical spine. J. Biomech. **39**(2), 375–380 (2006)

18. Bell, K.M., Debski, R.E., Sowa, G.A., Kang, J.D., Tashman, S.: (2019) Optimization of compressive loading parameters to mimic in vivo cervical spine kinematics in vitro. J. Biomech. **18**(87), 107–113 (2019)

19. Patel, V.V., Wuthrich, Z.R., McGilvray, K.C., Lafeur, M.C., Lindley, E.M., Sun, D., Puttlitz, C.M.: Cervical facet force analysis after disc replacement versus fusion. Clin. Biomech. (Bristol, Avon) **44**, 52–58 (2017)

20. Sun G., Li, S., Cao, Y., Lang, F.: Cervical Cancer diagnosis based on random forest. Int. J. Perform. Eng. **13**, 446–457. https://doi.org/10.23940/ijpe.17.04.p12.446457.

Influence of Compressive Preloading on Range of Motion and Endplate Stresses in the Cervical Spine During Flexion/Extension

Srikanth Srinivasan, R. Deepak, P. Yuvaraj, D. Davidson Jebaseelan,
Narayan Yoganandan, and S. Rajasekaran

1 Introduction

The clinical biomechanics of the spine can be characterized with the help of experimentally validated finite element (FE) models based on accurate geometry, material property, boundary and loading conditions [23]. FE models are useful in obtaining stresses and strains that are not commonly analysed in detail using in-vitro and in-vivo studies. FE models also provide an analytical method of understanding the underlying mechanisms of degeneration, leading to improved prevention and diagnosis of clinical problems [25].

The spine can maintain stability under various load conditions, however, in-vitro studies conducted by multiple groups reveal that cadaveric cervical spine specimens have critical buckling load between 10 and 15 N [13] as a result, most studies investigating multi-segment cervical spine have not included a compressive preload [12].

S. Srinivasan (✉) · D. Davidson Jebaseelan
School of Mechanical and Building Sciences, VIT University, Chennai Campus, India
e-mail: rohitvasan97@gmail.com

D. Davidson Jebaseelan
e-mail: davidson.jd@vit.ac.in

R. Deepak · P. Yuvaraj · N. Yoganandan
Center for Neurotrauma Research, Department of Neurosurgery, Medical College of Wisconsin,
Milwaukee, WI, USA
e-mail: drajasekaran@mcw.edu

P. Yuvaraj
e-mail: yuvapuru@mcw.edu

N. Yoganandan
e-mail: yoga@mcw.edu

S. Rajasekaran
Ganga Medical Center and Hospitals Pvt-Ltd, Coimbatore, India
e-mail: sr@gangahospital.com

© Springer Nature Switzerland AG 2021
C. T. Lim et al. (eds.), *17th International Conference on Biomedical Engineering*,
IFMBE Proceedings 79, https://doi.org/10.1007/978-3-030-62045-5_12

Patwardhan et al. [13–15] developed the "follower load" (FL) method of applying compressive preload to a multi-segment lumbar spine specimen without buckling, and adapted FL for application to the cervical spine. It was observed that in comparison to the hypermobility obtained in response to a compressive vertical load, the mechanical stability significantly increased on the application of a follower load [2]. This study aims to investigate the effect of such a load on the endplate stresses.

The vertebral end-plate is critical for maintaining the health of the intervertebral disc and plays an important role in the load transfer across the column. The endplates of a healthy disc prevent the highly hydrated nucleus from bulging into the adjacent vertebral bone, while simultaneously absorbing hydrostatic pressure that results from mechanical loading of the spine [11]. The objective of the present study is to fully investigate the biomechanical influence of a follower load in the cervical spine in terms of ROM, spinal kinematics and thereby on the endplate stresses (EPS) using a validated finite element model (FEM).

2   Methods

A compressive pre-load is applied to accurately simulate head mass and muscle forces, and is indicative of actual load transmission in the spine. An anatomically accurate and validated C2-T1 osteoligamentous model was used for this study [6, 19, 20]. The intact spinal model was compared with other finite element models as well as experimental corridors for validating ROM [4]. Material properties were obtained and modified within literature limits for the seven segments (C2-T1). Hexahedral elements were used to mesh the model (Table 1).

The model contained the five main soft tissues of the spine-anterior longitudinal ligament (ALL), posterior longitudinal ligament (PLL), capsular ligaments (CL), ligamentum flavum (LF), and interspinous ligaments (ISL). The complex behaviour of the ligaments was modelled using non-linear stress strain curves [9] and fabric membrane model. The intervertebral disc contained two distinct regions: nucleus pulposus and annulus fibrosus. The annulus consisted of annulus fibres modelled using non-linear stress strain curves and annulus ground substance, modelled as linear elastic. The posterior region consisted of four layers (total in 8), and the anterior region of the annulus fibrosus consisted of eight-pair layers (total in 16). The anterior annulus fibres formed a gap bilaterally at the uncovertebral clefts, and did not form a continuous ring with the posterior annulus fibres [10]. The number of elements on each level was 1392, 2060, 1970, 2060, 2130, and 1840, respectively from the C2–C3 to C7–T1 levels. The entire model (Fig. 1) contained a total of 11,452 elements for the annulus fibres.

External load of 2 Nm was applied at the center of gravity of the C2 vertebrae through a kinematic coupling. All degrees of freedom at the bottom endplate of T1 vertebrae was constrained. Compressive pre-loading of 100 N was applied using the FL methodology, with beam elements passing through the approximate center of rotation similar to in-vivo tests [1]. The same boundary and loading conditions were

Table 1 Material properties and models

Component	Element type	Constitutive model	Properties (Elastic modulus in Mpa)	Reference
Cortical bone	Hexahedral solid	Linear elastic	E = 16,800, μ = 0.3	Reilly and Burstein [18]
Trabecular bone	Hexahedral solid	Linear elastic	E = 400, μ = 0.3	Kopperdahl and Keaveny [7], Yoganandan et al. [24]
Annulus ground	Hexahedral solid	Linear elastic	E = 4.7, μ = 0.45	Kumaresan [8]
Nucleus pulposus	Hexahedral solid	Linear elastic	E = 1, μ = 0.49	Zhang [25]
Endplate	Quadrilateral shell	Linear elastic	E = 5600, μ = 0.3	Panzer and Cronin [14]
Facet cartilage	Quadrilateral shell	Linear elastic	E = 10, μ = 0.3	Yamada [21]
Ligaments	Quadrilateral mebrane	Non-linear fabric	Stress–strain curves	Mattucci et al. [9], Yoganandan et al. [22]

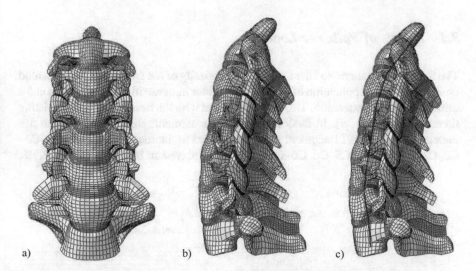

a)   b)   c)

Fig. 1 Intact model front view **a** side view **b** and preloading path **c**

applied to all finite element models. The solution was tuned to include the effects of large geometric deformations, material and contact non-linearities. ABAQUS (Simulia, Providence, RI, USA) was used to conduct the nonlinear analysis and post-processing.

The ROM was computed using a python code interfaced with ABAQUS. The code extracted nodal data from workspace and computed the deformed co-ordinates based on the simulation. The angle was then calculated using physiological planes between the deformed and undeformed coordinates.

3 Results

An exact three-dimensional model of the cervical spine was developed in this study. The model was simulated for sagittal moments of 2 Nm taking into account both material and geometric non-linearities. The results obtained for the intact model were first compared and validated with other finite element models [4] and experimental studies [19]. The most flexible segment in flexion was the C4–C5 segment, displaying markedly higher values of ROM compared to the other segments while the lower segments displayed almost equal flexibility. The C5–C6 segment was the most flexible segment in extension while the C2–C3 segment was stiffer compared to the rest of the model. The ROM for the intact model fell within literature limits for nearly all of the segments in both flexion and rotation as shown in Fig. 2.

3.1 Effects of Follower Load

The follower load increased the load-bearing capacity of the spine. The general trend observed after the application of the FL was a minor increase in ROM in flexion and a minor decrease in extension. This is in agreement with the trend reported by existing literature studies [1, 5]. In flexion, most of the segments displayed less than 5% increase, except C7–T1 segment which displayed an increase of 11.5%. The C2–C3, C3–C4, C4–C5, C5–C6, C6–C7 segments displayed an increase of 2.6%, 3.7%,

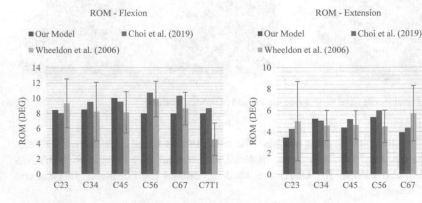

Fig. 2 ROM corridors for flexion and extension

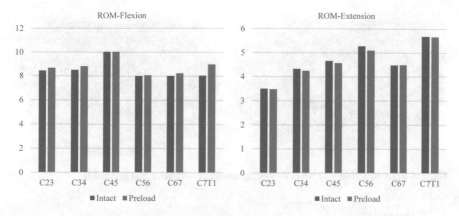

Fig. 3 ROM comparison between intact and preload for flexion and extension

0.01%, 0.7% and 2.6% in ROM respectively. In extension, all the segments exhibited less than 4% decrease in ROM, with C5–C6 showing the maximum decrease at 3.40%. The comparison of the ROM values for intact and preload model are shown in Fig. 3.

The endplate von Mises stresses for various physiological loadings varied between 2 and 10 MPa across the various functional units. In the case of pure moments, the stress contour obtained was asymmetric and seemed to be influenced by irregularities in geometry and curvature of the intact model (Fig. 4). However, after the application of a preload, the regions of maximum stress concentration seemed to be near the edges of the endplates in both flexion and extension. The stress distribution was also comparatively more symmetric (Fig. 5).

While the change in ROM is minimal, the results indicate that the application of a FL has an increased change in the location of localized stress in the endplates. Thus, the inclusion of a FL in modelling becomes necessary in Heterotopic Ossification (HO) studies, where the localisation of endplates stresses can possibly cause abnormal bone growth.

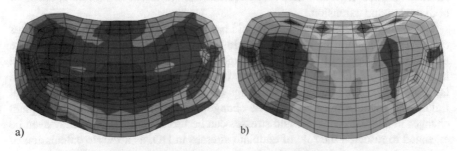

Fig. 4 C4 Superior Endplate stress contour in flexion—Intact **a** and Preload **b**

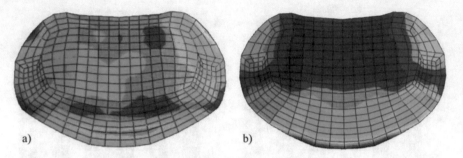

a)   b)

Fig. 5 C5 Superior Endplate stress contour in extension—Intact **a** and Preload **b**

4 Discussion

It was observed that the presence of a FL closely simulates in-vivo biomechanics. Closely approximating the testing conditions in computer models can lead to results with better accuracy, increasing the reliability and utility of internal response data sets with clinicians.

Recent studies (Barrey et al. 2012) [3] have documented the effect of a FL on ROM, however its effect on endplate stress has not been studied before. End plates have an important role in fluid flow, which in turn plays a major role in the mechanical behaviour of the disc and load transfer through the vertebral bodies. In the case of anterior cervical discectomy and fusion (ACDF), change in the angle between the endplates is indicative of device sinkage into the vertebral bodies. In order to evaluate the future success of implanted devices, in-depth knowledge of endplate morphology and biomechanics is necessary.

This work brought out the importance of studying the change in the cervical spine due to application of a preload, leading to changes in ROM and endplate stresses. It was seen that the stress in the endplates is distributed and maximum stress occurs towards the edges after the application of a pre-load, as opposed to without a pre-load where the stress distribution is more influenced by the curvature. Irrespective of the stress magnitude, regions where they occurred were different in both the models for both flexion and extension.

Removal of the cortical endplate could have a significant effect on the cancellous core stresses. Ideally the endplate should be left intact as much as possible. The FE analysis of devices like inter-body cages and artificial discs can provide a good insight regarding the load sharing at the interface of the fusion with the bone and is helpful to evaluate the critical loadings and cases which may lead to a potential failure and fracture at endplates. The change in ROM after ADR and subsequent changes in endplate behavior and stresses can be further studied. This may even be extended to research the role of endplate stresses in HO, as accurate calculation of endplate stresses and their location is necessary in order to predict possible regions of abnormal bone formation.

5 Conflict of Interest

The authors do not have any known conflicts of interest with regard to this publication.

Acknowledgements This study was supported by the Dassault Systemes Foundation, India; The Department of Neurosurgery, Medical College of Wisconsin, USA and Ganga hospitals, Coimbatore, India.

References

1. Barrey, C., Rousseau, M.A., Persohn, S., Campana, S., Perrin, G., Skalli, W.: Relevance of using a compressive preload in the cervical spine: an experimental and numerical simulating investigation. Eur. J. Orthop. Surg. Traumatol. **25**(Suppl 1), S155-165 (2015)
2. Barrey, C., Campana, S., Persohn, S., Perrin, G., Skalli, W. (2012). Cervical disc prosthesis versus arthrodesis using one-level, hybrid and two-level constructs: An in vitro investigation. *European Spine Journal, 21*(3), 432–442.
3. Bell, K.M., Debski, R.E., Sowa, G.A., Kang, J.D., Tashman, S.: Optimization of compressive loading parameters to mimic in vivo cervical spine kinematics in vitro. J. Biomech. **18**(87), 107–113 (2019)
4. Choi, H., Purushothaman, Y., Baisden, J, Yoganandan, N.: Unique biomechanical signatures of Bryan, Prodisc C, and Prestige LP cervical disc replacements: a finite element modelling study. Eur. Spine. J. (2019)
5. Du, C.F. et al.: The biomechanical response of cervical spine under different follower loads. In 2019 IEEE International Conference on Mechatronics and Automation (ICMA) pp. 360–364 (2019)
6. John, J.D., Saravana Kumar, G., Yoganandan, N.: Cervical spine morphology and ligament property variations: a finite element study of their influence on sagittal bending characteristics. J. Biomech. **85**, 18–26 (2019)
7. Kopperdahl, D.L., Keaveny, T.M.: Yield strain behavior of trabecular bone. J Biomech. **31**(7), 601–608 (1998)
8. Kumaresan, S., Yoganandan, N., Pintar, F.A., Maiman, D.J., Kuppa, S.: Biomechanical study of pediatric human cervical spine: a finite element approach. J. Biomech. Eng. **122**(1), 60–71 (2000)
9. Mattucci, S.F.E., Moulton, J.A., Chandrashekar, N., Cronin, D.S.: Strain rate dependent properties of younger human cervical spine ligaments. J. Mech. Behav. Biomed. Mater. **10**, 216–226 (2012)
10. Mercer, S., Bogduk, N.: The ligaments and annulus fibrosus of human adult cervical intervertebral discs. Spine (Phila Pa 1976) **24**(7), 619–626; discussion 627–618 (1999)
11. Moore, R.: The vertebral end-plate: what do we know? E. Spine J. **9**, 92 (2000)
12. Ng, H.W., Teo, E.C.: Influence of preload magnitudes and orientation angles on the cervical biomechanics: a finite element study. J. Spinal. Disord. Tech. **18**, 72–79 (2005)
13. Panjabi, M.M, Cholowicki, J., Nibu, K., et al.: Criticial load of the human cervical spine: an in vitro experimental study. Clin. Biomech. **13**, 11–17 (1998)
14. Panzer, M.B., Cronin, D.S.: C4–C5 segment finite element model development, validation, and load-sharing investigation. J. Biomech. **42**, 480–490 (2009)
15. Patwardhan, A.G., Meade, K.P., Lee, B.: A frontal plane model of the lumbar spine subjected to a follower load: implications for the role of muscles. J. Biomech. Eng. **123**, 212–217 (2001)
16. Patwardhan, A.G., Havey, R.M., Ghanayem, A.J., Diener, H., Meade, K.P., Dunlap, B., Hodges, S.D.: Load-carrying capacity of the human cervical spine in compression is increased under a follower load. Spine (Phila Pa 1976) **25**, 1548–1554 (2000)

17. Patwardhan, A.G., Havey, R.M., Meade, K.P., Lee, B., Dunlap, B.: A follower load increases the load-carrying capacity of the lumbar spine in compression. Spine (Phila Pa 1976) **24**, 1003–1009 (1999)
18. Reilly, D.T., Burstein, A.H.: The elastic and ultimate properties of compact bone tissue. J. Biomech. **8**(6), 393–405 (1975)
19. Wheeldon, J.A., Pintar, F.A., Knowles, S., Yoganandan, N.: Experimental flexion/extension data corridors for validation of finite element models of the young, normal cervical spine. J. Biomech. **39**(2), 375–380 (2006)
20. Wheeldon, J.A., Stemper, B.D., Yoganandan, N., Pintar, F.A.: Validation of a finite element model of the young normal lower cervical spine. Ann. Biomed. Eng. **36**(9), 1458–1469 (2008)
21. Yamada, H.: Strength of biological materials. Williams and Wilkins (1970)
22. Yoganandan, N., Kumaresan, S., Pintar, F.A.: Geometric and mechanical properties of human cervical spine ligaments. J. Biomech. Eng. **122**, 623 (2000)
23. Yoganandan, N., Kumaresan, S., Pintar, F.A.: 'Biomechanics of the cervical spine. Part 2. Cervical spine soft tissue responses and biomechanical modeling.' Clin. Biomech. **16**(1), 1–27 (2001)
24. Yoganandan N1, Pintar FA, Stemper BD, Baisden JL, Aktay R, Shender BS, Paskoff G, Laud P.: Trabecular bone density of male human cervical and lumbar vertebrae. Bone **39**(2), 336–44 (2006)
25. Zhang, Q.H., Teo, E.C., Ng, H.W., Lee, V.S.: Finite element analysis of moment–rotation relationships for human cervical spine. J. Biomech. **39**

Classification of Dementia MRI Images Using Hybrid Meta-Heuristic Optimization Techniques Based on Harmony Search Algorithm

N. Bharanidharan⊙ and Harikumar Rajaguru⊙

1 Introduction

Dementia is a collective term used to represent the decline of cognitive abilities. The demented subjects may have problems with memory, reasoning, decision making, etc. Around fifty million people are demented worldwide according to a survey conducted in 2017. The rate of dementia affected people is increasing drastically and it is estimated that this rate is doubling every 20 years [1–2]. Magnetic Resonance Imaging (MRI) is an efficient tool to examine the internal parts of human body and it has the ability to discriminate various soft tissues and grey matter present in the brain. So it is widely used diagnosing dementia [3]. Generally the clinician will diagnose the disease through visual inspection and sometimes it can be erroneous since large numbers of MRI images needs to be examined for each subject [4]. Hence a computerized classification technique will be useful for the clinician while diagnosis.

Hence the overall objective of this research work is to build an accurate classifier for categorizing the input images in two classes: Non-Demented (ND) and demented (DEM). Based on learning type, classification techniques are broadly classified into three types namely, supervised classification, unsupervised classification, and reinforcement learning. Our research work concentrates on unsupervised classification. In the unsupervised classification, the similarity between the input data points is analyzed and then it will group the data points into 'nc' number of classes. The value of 'nc' is user defined. The popular unsupervised classification approaches are K-Means clustering [5] and Fuzzy C Means (FCM) [6] clustering. These clustering

N. Bharanidharan (✉)
Vel Tech Rangarajan Dr.Sagunthala R & D Institute of Science and Technology, Chennai, India
e-mail: bharani2410@gmail.com

H. Rajaguru
Bannari Amman Institute of Technology, Sathyamangalam, India
e-mail: harikumarrajaguru@gmail.com

© Springer Nature Switzerland AG 2021
C. T. Lim et al. (eds.), *17th International Conference on Biomedical Engineering*,
IFMBE Proceedings 79, https://doi.org/10.1007/978-3-030-62045-5_13

techniques are based on the following principle: Minimize the intra-class distance and maximize the inter-class distance.

Nowadays meta-heuristic techniques are very popular due to their easiness, flexibility, derivation free mechanism. Usually meta-heuristic techniques are built using the inspiration from characteristics of animals, physical phenomena or evolutionary concepts [7]. Some examples of meta-heuristic algorithms include Genetic Algorithm (GA), Particle Swarm Optimization (PSO), Artificial Bee Colony (ABC), Ant Colony Optimization (ACO), Harmony Search (HS), etc. To improve the performance, hybridization of two different techniques is usually followed. In literature, reports can be found regarding the usage of HS based hybrid techniques which are applied for solving numerical optimization problem, selection of features for a classifier, and training neural networks. In addition, usage of meta-heuristic techniques to solve clustering problem is also common. But it is ingenious to use HS based hybrid meta-heuristic algorithms as transformation technique based classification algorithm. As a transformation technique, HS based hybrid meta-heuristic algorithms will try to convert non-linearly separable features into linearly separable features.

The remaining part of the paper is organized as follows: overall methodology used in this paper is outlined in second section and basics of meta-heuristic techniques are presented in the third section. Fourth and fifth section presents the procedure for implementing the individual and hybrid classifiers respectively. The results are discussed in the sixth section and concluded in the last section.

2 Overall Methodology

This research work proposes HS based three different hybrid meta-heuristic algorithms namely HS-PSO, HS-ABC, and HS-ACO as transformation technique based classifiers. To prove the better performance provided by HS based hybrid methods, FCM based semi meta-heuristic hybrid methods namely FCM-PSO, FCM-ABC, and FCM-ACO are used. In addition, efficiency of HS based hybrid methods are established through the comparison with popular unsupervised classification approaches like K-Means, FCM, and standard clustering approach implementation using PSO, ABC, ACO and HS [8]. The flowchart depicting the overall methodology used in this research work is shown in Fig. 1.

The individual classifier implementation can be explained as follows: The input MRI image is divided into 'n' equal regions and then five statistical features namely mean, variance, skewness, kurtosis, and entropy are computed for each region. These 80 features are given as input to the classifier. Based on the 'n' value, there is trade-off between classification accuracy and execution time. Small value of 'n' may leads to less inaccurate classification and large value of 'n' may need high computational time. As depicted in Fig. 2, the ideal value for 'n' is computed as 16 through experiments for all the classifiers used in this research work.

The hybrid classifier implementation can be explained as follows: The size of the input MRI brain image is 208*176 and so there will 36,608 number of intensity

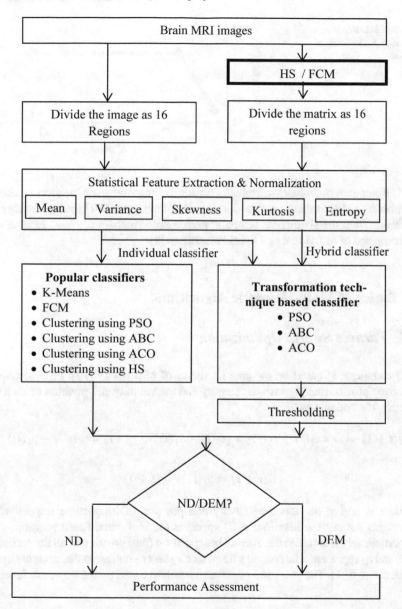

Fig. 1 Flowchart of overall methodology used in this research work

values. These intensity values are directly assigned as input to HS or FCM technique. Then the output matrix given by HS or FCM is divided into 16 equal regions and 5 features are extracted in each region. This process will results in 80 features and this will be given as input to the popular classifiers or transformation technique based classifier.

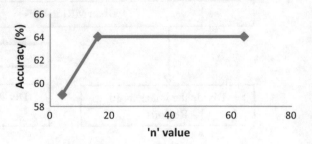

Fig. 2 Classification accuracy for various 'n' values

The output of meta-heuristic transformation technique is given to automatic binary thresholding algorithm to take the decision about the input MRI image as either ND or DEM. Then the six different popular performance metrics are used to assess the performance of individual and hybrid classifiers [9].

3 Basics of Meta-Heuristic Algorithms

3.1 Particle Swarm Optimization

PSO technique is based on the characteristics of bird flocks [10]. PSO belongs to iterative optimization algorithms category and the velocity and position of each bird (particle) is updated as

$$v_i(t + 1) = w * v_i(t) + c_1 * r_1 * (p_i(t) - x_i(t)) + c_2 * r_2 * (gbest - x_i(t)) \quad (1)$$

$$x_i(t + 1) = x_i(t) + v_i(t + 1) \quad (2)$$

Here v_i and x_i denotes the velocity and position of ith particle respectively; t represents the current iteration; $p_i(t)$ specifies the individual finest position of the ith particle; *gbest* signifies the over-all best position (best position of all the particles); w, c_1 and c_2 represents the control parameters; r_1 and r_2 represent the random number in the range [0,1]. The $p_i(t)$ and *gbest* are identified based on the objective function.

3.2 Artificial Bee Colony

ABC technique is based on the food searching mechanism of bees [11]. The position of food sources will be stored in the memory of bees and distinction of food sources (x_i) is quantified by its nectar level. To validate the quality of food sources, fitness value ($f(x_i)$) will be used. Selection methods like roulette wheel selection is used to identify the food bases and this process will depend on probability p_i which is computed as,

$$p_i = \frac{f(x_i)}{\sum_{i=1}^{n} f(x_i)} \tag{3}$$

Then the new position of selected ith food source (x_i') will be updated iteratively as,

$$x_i' = x_i + \alpha(x_i - x_k) \tag{4}$$

Here x_i refers to the old position of ith food source and k is selected randomly to represent neighbor solution; α denotes to the control parameter in the range [-1, + 1]. If the fitness function of the new solution is greater than the previous one, the new solution will be retained otherwise old solution will be used.

3.3 Ant Colony Optimization

ACO is based on the food searching manner followed by ants [12]. Depending upon the values of trial and attractiveness, the travel of each ant will be decided in this iterative algorithm. The movement probability of kth ant (p_{ij}^k), to transfer from position i to j will be given as,

$$p_{ij}^k = \begin{cases} \frac{\omega_{ij}^\alpha + \epsilon_{ij}^\beta}{\sum_{(k)\epsilon allowed_k} \left(\omega_{ij}^\alpha + \epsilon_{ij}^\beta \right)} & if \ j \ \epsilon \ allowed_k \\ 0 & Otherwise \end{cases} \tag{5}$$

In the above equation, ϵ_{ij} denotes the attractiveness which is inversely proportional to the distance between i and j; α and β are the control parameters in the range [0,1]. The trials present in the path i to j will be iteratively updated as,

$$\omega_{ij}(t + 1) = \rho.\omega_{ij}(t) + \Delta\omega_{ij} \tag{6}$$

where, ρ is the control parameter in the range [0,1] which affects the influence of previous trial ($\omega_{ij}(t)$) on the updated value ($\omega_{ij}(t + 1)$). The adjustment of the trial value ($\Delta\omega_{ij}$) from i to j is calculated as,

$$\Delta\omega_{ij} = \sum_{k=1}^{m} \Delta\omega_{ij}^k \tag{7}$$

$$\Delta\omega_{ij}^k = \begin{cases} \frac{Q}{L_k} & if\,k-than\,tuses\,the\,path\,ito\,j \\ 0 & otherwise \end{cases} \tag{8}$$

Here Q represents a constant and L_k denotes the total distance covered by ant k.

3.4 Harmony Search Algorithm

HS is inspired by the enhancement process carried out by musicians during music composition or playing [13–14]. The process of tuning the instruments for getting better music is termed as the enhancement process.

Algorithm 1: Harmony Search Algorithm

1:    Initialize HMS, HMCR, PAR, BW, MAXIT, and Harmony Memory (x)

2:    **Repeat**

3:      **For** (I = 1:HMS)

4:      **For** (J = 1:N)

5:       **If** (r_1 < HMCR)

6:        $x_{IJ\,(new)} = x_{IJ}$

7:        **If** (r_2 < PAR)

8:         $x_{IJ\,(new)} = x_{IJ\,(new)} + r_3 * BW$

9:        **end if**

10:        **If** ($x_{IJ\,(new)} < x_{I\,L}$)

11:         $x_{IJ\,(new)} = x_{I\,L}$

12:        **end if**

13:        **If** ($x_{IJ\,(new)} > x_{I\,U}$)

14:         $x_{IJ\,(new)} = x_{I\,U}$

15:        **end if**

16:       **else**

17:        $x_{IJ(new)} = x_{I\,L} + r_4 * (x_{I\,U} - x_{I\,L})$

18:       **end if**

19:      **end for**

20:      **end for**

21:      **If** ($f(x_{new}) < f(x_{old})$)

22:           update harmony memory with x_{new}

23:     **end if**

24:     set $t = t + 1$

25: **until** (MAXIT reached)

26: Get the best solution x_{new}

The algorithm for HS is presented in Algorithm 1 and the following parameters are initialized: Harmony Memory Size (HMS), Harmony Memory Considering Rate (HMCR), Pitch Adjusting Rate (PAR), Band Width (BW), and Maximum number of iterations (Maxit). In Algorithm 1, $f(x_{new})$ and $f(x_{old})$ denotes the fitness of new and old harmony memory respectively; N represents the number of solutions; r_1, r_2, r_3, and r_4 denotes the random number in the range [0, 1]; Lower bound is considered as 0 and upper bound is considered as 1.

4 Implementation of Individual Classifiers

The procedure to implement PSO as transformation technique based individual classifier can be described as follows: In general, if PSO is used for feature selection and solving optimization problem, the particles will be initialized randomly. But to solve classification problem, the 80 particles are initialized with 80 normalized statistical features. The upper bound of normalized value will be equal to 1 and that will be used as target. Fitness function relates to the reciprocal of Euclidean distance from target to the current particle's position. Based on this fitness values, global best and personal best are computed and then Eqs. (1 and 2) are used to update the position of particles in each iteration.

Three control parameters namely w, c_1, and c_2 are existing in PSO. In general, the range of w is [0, 1] and the range of both c_1 & c_2 will be [0, 4] as specified in [15]. Hence these three hyper-parameters are initialized at their mid-points: $w = 0.5$, $c_1 = 2$ and $c_2 = 2$ while finding the ideal value for maxit using Trial & Error (TE) method as depicted in Fig. 3.

The classification accuracy for various values of maxit is tested as depicted in Fig. 3. The ideal value for maxit is found as 3 for which the highest accuracy of 62% is achieved. As a next step, the optimum values for PSO hyper-parameters (w, c_1 & c_2) have to be found. Usually w will have more influence on the classification performance when compared to $c1$ & $c2$. So c_1 & c_2 will be fixed at its mid-point (0.5) and as depicted in Fig. 4, the ideal value for w is found as 0.9 using TE method. For this ideal w value, highest accuracy of 64% is achieved.

With the ideal values for *maxit* and w, the ideal values for c_1 and c_2 can be found simultaneously using TE method. As depicted in Fig. 5, the best values for c_1 and c_2 are found as 0.1 for which the highest accuracy of 65% is possible in the statistical features usage case. In total, the ideal values are maxit $= 3$, w $= 0.9$, c1 $= 0.1$ and c2 $= 0.1$. In the similar manner, the other techniques like ABC, ACO, and HS are

Fig. 3 Finding the ideal
value for maxit in PSO

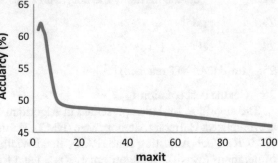

Fig. 4 Finding the ideal
value for inertia weight (w)
in PSO

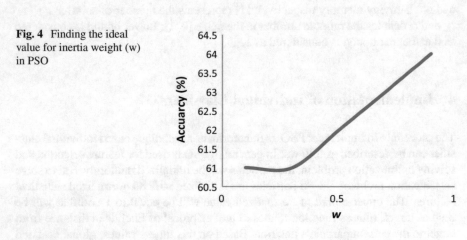

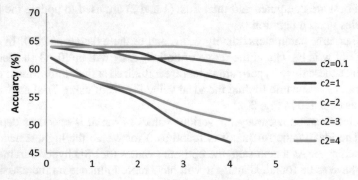

Fig. 5 Finding ideal values for c_1 and c_2 in PSO

implemented as transformation technique based classifier individually and the ideal
values found using TE method for these techniques are presented in Table 1.

Table 1 Ideal value for control parameters present in various meta-heuristic techniques using TE method

SI technique	Control parameter	Ideal Values
PSO	W	0.9
	c_1	0.1
	c_2	0.1
ABC	A	0.5
ACO	ρ	0.7
	α	0.5
	β	0.7
HS	HMCR	0.9
	PAR	0.1

5 Implementation of Hybrid Classifiers

5.1 HS based hybrid classifier

HS has a unique feature of iterating the solutions within a bound. This feature is not present in other meta-heuristic techniques like PSO, ABC and ACO. In other words, the data points will present only in the middle of upper and lower bounds in HS. This unique ability of HS will be useful to enhance the performance of individual classifiers.

The dataset contains eight bit grey scale images and the grey value will be in the range of [0, 255]. The whole image of dimension 208*176 is assigned as input to the HS and so the 36,608 elements are used to initialize the solutions in harmony memory. The ideal values found for individual classifiers (given in Table 1) are used even in hybrid classifiers. Once the HS reaches the maxit, the output matrix of iterative HS algorithm will be divided equally into 16 sub-matrices and 5 statistical features are extracted from each region. This process will result in 80 feature values and they will be used to initialize the solutions of any one of the following technique: PSO, ABC, or ACO.

The selection of value for upper and lower bounds is critical to get the improved accuracy. If the difference between lower and upper bounds is very less, the solutions are reaching the identical points with small variance; when the difference is too big, then the classification accuracy is getting reduced. So the difference should present ideally around the mean value 55. To divide the intensity range easily into five bands, 51 is selected. The 5 various bands are (0–51), (51–102), (102–153), (153–204), (204–255) and the accuracy values for different bands are shown in Fig. 6. Through Fig. 6, the ideal band is found as (204–255) on which the highest accuracy is achieved for three different HS based hybrid algorithms.

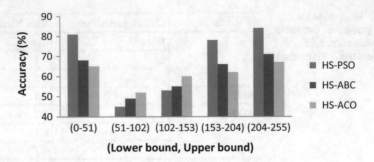

Fig. 6 Selection of ideal values for upper and lower bounds in HS based hybrid classifier

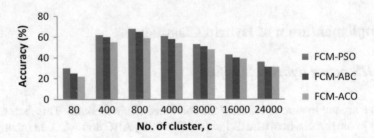

Fig. 7 Finding the ideal number of clusters for FCM based hybrid algorithms

5.2 FCM Based Hybrid Classifier

The centroid of clusters is used are computed using FCM and statistical features are calculated from the output of FCM and these features are used to initialize the solutions of the meta-heuristic technique which has to be improved.

Initialization of the feature vector x_k is done through 36,608 grey values. Number of clusters, c have more influence on the performance of FCM and so the ideal value of c is found as 800 at which the better performance is yielded for all the three FCM based hybrid algorithms as depicted in Fig. 7. FCM computes 800 cluster centroids for the input MRI image and these values are divided into sixteen sub-matrices. Five statistical features are computed for sixteen sub-matrices and the resultant eighty statistical features are used to initialize the solutions of PSO/ABC/ACO.

6 Results and Discussion

Using the procedure described in fourth section, the individual meta-heuristic techniques like PSO, ABC, ACO, and HS are implemented as transformation technique based classifier and six different performance metrics namely Accuracy (ACC), Error Rate (ER), Sensitivity (SENS), Specificity (SPEC), False Positive Rate (FPR), and

Precision (PREC) are computed for each classifier and presented in Table 2. Similarly HS and FCM based hybrid classifiers are implemented using the procedure described in fifth section and the performance metrics given by the proposed classifiers are given in Table 3.

The accuracy yielded by various classifiers is depicted in Fig. 8. When compared to popular K-Means and FCM techniques, transformation technique based classifiers

Table 2 Classification performance metrics of various individual classifiers

	ACC	ER	SENS	SPEC	FPR	PREC
PSO	66	34	62	69	31	62
ABC	61	39	46	72	28	57
ACO	56	44	69	46	54	51
HS	62	38	63	60	40	56
KM	59	41	63	55	45	53
FCM	58	42	62	55	45	52
PSO CLUS	61	39	63	58	42	55
ABC CLUS	57	43	62	54	46	52
ACO CLUS	58	42	67	51	49	52
HS CLUS	60	40	63	57	43	54

Table 3 Classification performance metrics of various hybrid classifiers

	ACC	ER	SENS	SPEC	FPR	PREC
HS-PSO	84	16	85	83	17	80
HS-ABC	71	29	69	72	28	67
HS-ACO	67	33	73	62	38	60
FCM-PSO	68	32	67	69	31	64
FCM-ABC	66	34	77	57	43	59
FCM-ACO	60	40	69	52	48	54

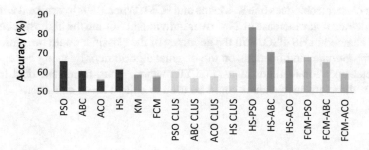

Fig. 8 Accuracy comparison of various individual and hybrid classifiers

like PSO, ABC, and HS are performing well. In addition the transformation technique based classifiers outperforms the commonly used clustering approaches done through PSO, ABC, ACO and HS. The reason for this significant performance can be established through investigating the following points: The vital difference between standard clustering approaches and transformation based classifiers is the objective function. In the standard clustering approach, the objective function is minimized for reducing the intra-class distance. In the transformation based classification technique, Euclidean distance between data points and target is used as fitness function. Based on the fitness value, each data point is iteratively updated. The classification accuracy differs in each iteration and highest accuracy is attained at particular iteration; that iteration umber is considered as ideal value for maxit. This process is carried out for all the SI algorithms. For example, in PSO implementation, this process is represented in Fig. 3 and this process is the key for success behind the transformation technique based classifiers using meta-heuristic algorithms.

When HS is hybridized with PSO, it provides accuracy increase of 27% over individual PSO while 16–18% accuracy increase is witnessed in HS-ABC and HS-ACO when compared to individual ABC and ACO. The FCM based hybrid algorithms are giving comparatively low performance and their accuracy improvement is just 2–11%. Remarkably, HS-PSO gives the highest accuracy of 84% for dementia classification. The reason for success of HS-PSO can be explained as follows: PSO is week in exploration but good in exploitation and HS helps to improve the exploration capability through hybridization. In addition, the transformation capability of HS to convert non-linearly separable direct intensity values into linearly separable values helps the PSO to attain global optima.

7 Conclusion

This research work analyzes the performance enhancement offered by hybridizing HS with other meta-heuristic algorithms like PSO, ABC, and ACO in transformation technique based dementia classification. The results clearly show the significant performance of HS based hybrid algorithms over FCM based hybrid algorithms and popular cluster techniques like K-Means and FCM. When HS is hybridized with PSO, it provides accuracy increase of 27% over individual PSO and the highest accuracy of 84% is attained in HS-PSO. Still the accuracy of the classifier could be improved by using appropriate modifications on the original algorithm or by using better control parameter updation methods instead of TE method. Apart from statistical features, age, sex, direct intensity values could be used to build a better classifier.

References

1. International statistical classification of diseases and related health problems, 10th Revision. Geneva, World Health Organization (1992)
2. Paraskevaidi, M., Martin-Hirsch, P.L., Martin, F.L.: Progress and challenges in the diagnosis of Dementia: a critical review. ACS Chemical Neuroscience 9(3), 446–461 (2018). https://doi.org/10.1021/acschemneuro.8b00007
3. Murray, A.D.: Imaging approaches for Dementia. American J. Neuroradiology. https://doi.org/10.3174/ajnr.A2782 (2011)
4. Kapse, R.S., Salankar, S.S., Babar, M.: Literature survey on detection of brain tumor from MRI Images. IOSR J. Electron. Commun. Eng. (IOSR-JECE) 10(1), 80–86 (2015)
5. Taufik, A., Syed Ahmad, S.S.: A comparative study of fuzzy C-Means And K-Means clustering techniques. 8th MUCET, Melaka, Malaysia (2014)
6. Parker, J.K., Hall, L.O.: Accelerating Fuzzy-C means using an estimated subsample size. IEEE Trans Fuzzy Syst 22(5) (2014)
7. Mirjalili, S., Mirjalili, S.M., Lewis, A.: Grey wolf optimizer. Adv. Eng. Softw. 69, 46–61 (2014)
8. Mahamed Omran, Andries Engelbrecht, Ayed A. Salman, "Particle swarm optimization method for image clustering," International Journal of Pattern Recognition and Artificial Intelligence, DOI: 10.1142, 2005
9. Fielding, A.H.: Cluster and Classification Techniques for the Biosciences. Cambridge University Press, Cambridge (2006)
10. Kennedy, J., Eberhart, R.: Particle swarm optimization. IEEE Int. Conf. Neural Netw. Australia 4, 1942–1948 (1995)
11. Karaboga, D.: An idea based on honey bee swarm for numerical optimization," Technical Report-TR06. Erciyes University, Engineering Faculty, Computer Engineering Department (2005)
12. Duan, H., Yu, X.: Hybrid ant colony optimization using memetic algorithm for traveling salesman problem. In Proceedings of the IEEE Symposium on Approximate Dynamic Programming and Reinforcement Learning (2007)
13. Geem, Z., Kim, J., Loganathan, G.V.: A new heuristic optimization algorithm. Harmony Search Simulation. 76, 60–68 (2001)
14. Omran, M.G.H., Mahdavi, M.: Global-best harmony search. Appl. Math. Comput. 198, 643–656 (2008)
15. Premalatha, K., Natarajan, A.M.: Hybrid PSO and GA models for document clustering. Int. J. Adv. Soft Comput. Appl. 2(3) (2010)

Classification of B-Cell Acute Lymphoblastic Leukemia Microscopic Images Using Crow Search Algorithm

N. Bharanidharan⬡ and Harikumar Rajaguru⬡

1 Introduction

Acute Lymphoblastic Leukemia (ALL) is a kind of blood cancer and usually elastic tissue like structure inside the bone where platelets are present will get affected. According to a survey, nearly 876,000 individuals' suffer ALL in 2015 and it leads to around 111,000 deaths in 2015 [1–3]. Blood Microscopic image captured using the digital microscope helps in diagnosis of this disease and it will be tedious task for the clinician to visually inspect the large number of microscopic images for each subject. Hence an automated classification technique will be helpful to the clinician during diagnosis of ALL [4–6]. So the objective of this research work is to classify blood smear microscopic images into two classes: Leukemic B-Cell Acute Lymphoblast Leukemia cells (ALL) and normal B-lymphoid precursors (HEM).

Nowadays meta-heuristic approaches are widely used in various applications due to their easiness, flexibility, derivation free mechanism. Usually meta-heuristic techniques are built using the inspiration from characteristics of animals, physical phenomena or evolutionary concepts. Swarm Intelligence (SI) techniques falls under meta-heuristic techniques category and SI algorithm are formed through the inspiration from group of animals [7, 8]. Some examples of SI algorithms include Particle Swarm Optimization (PSO), Artificial Bee Colony (ABC), Crow Search algorithm (CSA), etc. In literature, reports can be found regarding the usage of CSA for solving numerical optimization problem, selection of features for a classifier, and training neural networks. In addition, usage of SI algorithms to solve clustering problem is also common. But it is ingenious to use CSA as transformation technique based

N. Bharanidharan (✉)
Vel Tech Rangarajan Dr. Sagunthala R&D Institute of Science and Technology, Chennai, India
e-mail: bharani2410@gmail.com

H. Rajaguru
Bannari Amman Institute of Technology, Sathyamangalam, India
e-mail: harikumarrajaguru@gmail.com

© Springer Nature Switzerland AG 2021
C. T. Lim et al. (eds.), *17th International Conference on Biomedical Engineering*,
IFMBE Proceedings 79, https://doi.org/10.1007/978-3-030-62045-5_14

classification algorithm to convert non-linearly distinguishable points into linearly distinguishable points.

Outline of the remaining article: overall methodology used in this paper is given in second section and the third section presents the basics of CSA. Next section presents the procedure for implementing the CSA based classifier. The outcomes are deliberated in the fifth section and conclusion is presented in the last section.

2 Overall Methodology

The flowchart depicting the methodology implemented in this study is shown in Fig. 1. Two different cases are considered in this analysis namely with statistical

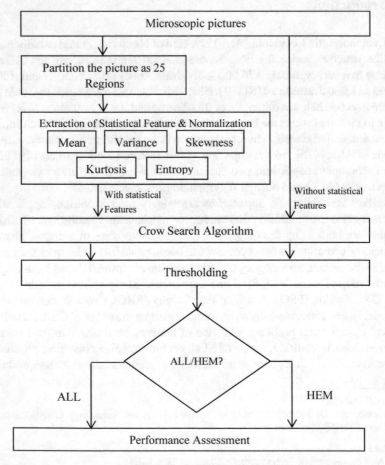

Fig. 1 Flowchart depicting the methodology implemented in this study

Table 1 Confusion matrix for the Leukemia microscopic image classification problem

Confusion matrix	Class	Predicted	
		HEM	ALL
Actual	HEM	TN	FP
	ALL	FN	TP

features (WS) and without statistical features (WOS). In WOS case, the direct intensity values from the microscopic images will be given as input to the CSA while in WS case, the microscopic image is divided into 25 regions and then five statistical features namely mean, variance, skewness, kurtosis and entropy are calculated for all region. These computed statistical features will be given as input to CSA in WS case. CSA will iterate the solutions and stops when the stopping criterion is met. Then a binary thresholding algorithm is used to determine the class of the given microscopic images as either HEM or ALL. From these results, confusion matrix is built as shown in Table 1 and then ten popular performance metrics [9] are used to calculate the excellence of classifier.

True Positive (TP) denotes sum of ALL images appropriately categorized as ALL while True Negative (TN) represents sum of HEM images appropriately categorized as HEM in Table 1; False Positive (FP) represents the total number of HEM images which are misclassified as ALL while False Negative (FN) denotes sum of ALL images which are inappropriately classified as HEM.

Ten different popular performance metrics used are given below:

The total correctness of the classifier,

$$\text{Accuracy}(\text{ACC}) = \frac{\text{TN} + \text{TP}}{\text{TN} + \text{TP} + \text{FN} + \text{FP}} \times 100\% \qquad (1)$$

Comprehensive amount of wrong classification will be,

$$\text{ErrorRate}(\text{ER}) = \frac{\text{FN} + \text{FP}}{\text{TN} + \text{TP} + \text{FN} + \text{FP}} \times 100\% \qquad (2)$$

The percentage of correctly classifying DEM images under DEM class will be,

$$\text{Sensitivity}(\text{SENS}) = \frac{\text{TP}}{\text{TP} + \text{FN}} \times 100\% \qquad (3)$$

The percentage of correctly classifying ND images under ND class will be,

$$\text{Specificity}(\text{SPEC}) = \frac{\text{TN}}{\text{TN} + \text{FP}} \times 100\% \qquad (4)$$

The percentage of wrongly classifying ND images as DEM,

$$FalsePositiveRate(FPR) = \frac{FP}{TN + FP} \times 100\% \qquad (5)$$

The percentage of perfectly classifying DEM images as DEM,

$$\textbf{Precision(PREC)} = \frac{\textbf{TP}}{\textbf{TP + FP}} \times 100\% \qquad (6)$$

The harmonic mean of precision & sensitivity,

$$\textbf{F1score} = \frac{\textbf{2TP}}{\textbf{2TP + FP + FN}} \times 100\% \qquad (7)$$

The correlation among actual and forecast is given by Mathews Correlation Coefficient (MCC),

$$\textbf{MCC} = \frac{\textbf{TPXTN - FPXFN}}{\sqrt{(TP + FP)(TP + FN)(TN + FP)(TN + FN)}} \times 100\% \qquad (8)$$

Jacard Metric (JM) openly ignores the perfect classification rate of true samples,

$$\textbf{JM} = \frac{\textbf{TP}}{\textbf{TP + FN + FP}} \times 100\% \qquad (9)$$

Balanced Classification Rate (BCR) will be very much useful in imbalanced datasets and it will be given as,

$$\textbf{BCR} = \frac{1}{2} \left(\frac{\textbf{TP}}{\textbf{TP + FN}} + \frac{\textbf{TN}}{\textbf{TN + FP}} \right) \times 100\% \qquad (10)$$

The reason for selecting 25 number of regions can be explained as follows: CSA is a population centered technique and so a group of crow has to be considered. If the microscopic image is not divided into regions, then only five data points produced by above mentioned five features will be there and this is not sufficient to initialize CSA. Hence the input microscopic picture is partitioned as 'n' identical regions & then five statistical features are calculated for each region. This 'n' number of features is given as input to the classifier. Based on the 'n' value, there is trade-off between classification accuracy and execution time. Small value of 'n' may leads to less inaccurate classification and large value of 'n' may need high computational time. As represented in Fig. 2, ideal value for 'n' is computed as 25 through experiments.

Histogram is widely used to investigate the linear discrimination of features that belongs to two different classes. If the features of two cases (ALL &HEM) are linearly separable then the classification of microscopic image can be done using simple thresholding technique. The difficulty involved in classification can be inferred based on valley between two classes. So as to investigate the linearity, the five statistical features for all the microscopic images considered in this analysis are computed and

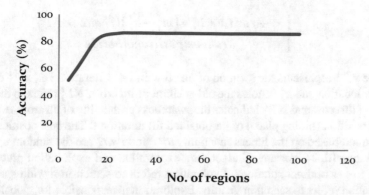

Fig. 2 Classification accuracy for various 'n' values

Fig. 3 Histogram of statistical normalized features

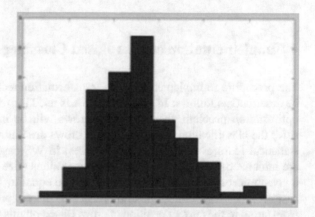

normalized to make the histogram plot. Unfortunately, these features are non-linear in nature as shown in Fig. 3. From the histogram plot, absence of valley can be witnessed which symbolizes the non-linearity of features. Hence there is a need of an algorithm better than thresholding and it should be able to classify the images accurately.

3 Basics of Crow Search Algorithm

CSA is formed through the inspiration from the intelligent behavior of crows. Usually a crow will try to steal the resources available in the hiding places of another crow [9]. This behavior is modelled as iterative CSA algorithm and position of crows are updated using the equation,

$$x_i^{t+1} = \begin{cases} x_i^t + ran1 * FL_i^t * \left(m_j^t - x_i^t\right) if ran2 > AP_j^t \\ a\, random\, position\, otherise \end{cases} \tag{11}$$

Here x_i^{t+1} represents the location of ith crow in t + 1 iteration i.e., x_i^{t+1} denotes the new location and x_i^t denotes the old position of ith crow. FL_i^t refers to the flight length of ith crow and AP_j^t indicates the awareness probability of jth crow. m_j^t refers to best position (hiding place) of jth obtained till iteration t. The best position of jth is identified based on the fitness function. $ran1$ and $ran2$ are the random numbers in the range [0, 1]. For any SI algorithm, exploration and exploitation process are crucial to get optimal solutions. Exploration refers to search in a wide space and exploitation refers to search in vicinity. Exploration is responsible for global search and avoiding local optima problem. In CSA, if the value of FL is very small, then it will be good in exploitation and if it is large, then it will be good in exploration [10].

4 Implementation of CSA Based Classifier

The procedure to implement CSA as transformation technique based classifier can be described as follows: In general, if CSA is used for feature selection and solving optimization problem, the position of crows will be initialized randomly. But to solve the classification problem, the 125 crows are initialized with 125 normalized statistical features (25 regions * 5 features) in WS case while the 202,500 crows are initialized with 202,500 direct intensity values (size of the image: 450*450) of a given microscopic image. The target will be equal to 1 and 300 in WS and WOS cases respectively. Fitness function relates to the reciprocal of Euclidean distance from target to the current position of crow. Based on this fitness values, best position (hiding place) of each crow will be determined and then Eq. (11) is used to update the position of position of each crow iteratively.

Two control parameters namely AP and FL exist in CSA. In general, the range of AP is [0, 1] and the range of FL is [0, 2] [11]. Hence these three hyper-parameters are initialized at their mid-points: AP = 0.5, and FL = 1 while finding the ideal value for Maximum number of iterations (MAXIT) through Trial & Error (TE) approach depicted in Fig. 4.

The highest accuracy of 76% is attained when MAXIT = 7 in WS case while the highest accuracy of 84% is attained when MAXIT = 10 in WOS case and this can be witnessed from the Fig. 4. Next the ideal values of AP and FL are found by keeping the optimum values for MAXIT as shown in Figs. 5 and 6 for WS and WOS cases respectively.

Through Fig. 5, the ideal values of both FL and AP are found as 0.5 at which the highest accuracy of 79% is achieved. Similarly through Fig. 6, the ideal values of FL and AP are found as 1.3 and 0.5 respectively at which highest accuracy of 87% is achieved.

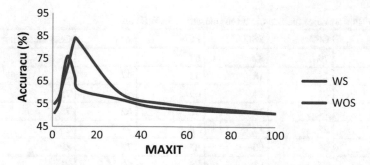

Fig. 4 Selection of ideal value for MAXIT

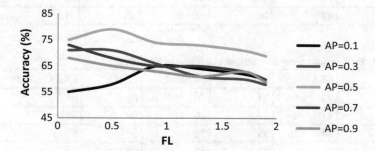

Fig. 5 Selection of ideal values for AP and FL in WS case

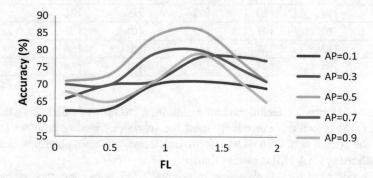

Fig. 6 Selection of ideal values for FL and AP in WOS case

5 Results and Discussion

The performance metrics obtained by using CSA as transformation technique based classifier in WS and WOS cases are presented in Tables 2 and 3. In addition, popular unsupervised classification techniques like FCM [12] and K-Means (KM) [13] are used to establish the significant performance of CSA. To prove the superiority of

Table 2 Performance metrics of various classifiers in WS case

	CSA	PSO	FCM	KM	CSA clustering
ACC	79	82	67	53	75
ER	21	18	33	47	25
SENS	81	84	69	49	75
SPEC	72	77	62	62	75
FPR	28	23	38	38	25
PREC	88	90	83	77	89
F1	85	87	75	60	81
MCC	50	57	28	10	45
JM	73	77	61	43	68
BCR	77	80	66	56	75

Table 3 Performance metrics of various classifiers in WOS case

	CSA	PSO	FCM	KM	CSA Clustering
ACC	87	73	70	60	65
ER	13	27	30	40	35
SENS	89	75	73	58	61
SPEC	80	69	63	66	75
FPR	20	31	37	34	25
PREC	92	86	84	82	86
F1	91	80	78	68	71
MCC	67	41	33	21	32
JM	83	67	64	51	56
BCR	85	72	68	62	68

CSA as transformation technique based classifier, PSO is evaluated using the same approach. SI algorithms are widely used for solving clustering problem [14]. To clarify the goodness of CSA as transformation technique based classifier, clustering approach using CSA [15] is also evaluated.

The accuracy of various classifiers in WS and WOS cases are compared in Fig. 7.

Transformation technique based classifier using CSA provides the better accuracy when compared to all other classification approaches used. Apart from the nature of equations used in the classifier, the main reason for success of this technique will be the usage of ideal values for MAXIT, AP and FL. CSA is providing better accuracy of 87% in WOS case and the accuracy is 79% in WOS case. The reason for better performance in WOS case over WS case can be stated as follows: In WS case, only 80 data points are used to initialize the swarm and so population of swarm is very small; In WOS case, 36,608 data points are used to initialize the swarm and so population of swarm is very large. But usage of direct intensity values provides more

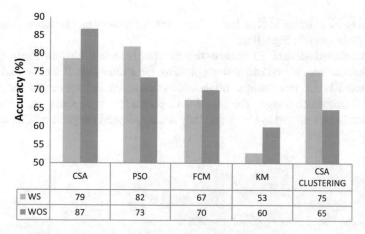

Fig. 7 Accuracy comparison of various classifiers in WS and WOS cases

information than the statistical features. In order to search in large solution space, the SI algorithm should have better exploration capability. CSA has good equilibrium among exploration & exploitation due to the presence of FL control parameter. The ideal value of FL is 0.5 and 1.3 in WS and WOS cases respectively. In WS case, lesser FL courtesies more exploitation and large value of FL favors more exploration. Due to this ability, CSA uses the virtue of having more information in WOS case appropriately and it is capable of producing good performance metrics.

PSO is giving better performance in WS case but comparatively poor performance in WOS case. Many articles in literature report the better exploitation capability of PSO and weakness of PSO in exploration. Hence PSO works well when number of input data points is less but lags to search globally when there are large number of input data points.

CSA clustering technique offers poor performance when compared to CSA in transformation technique based classifier. The reason for this can be understood by analyzing the following points: Clustering is one of the popular techniques used to implement unsupervised classification. Usually in clustering, intra-class distance has to be reduced or inter-class distance has to be increased. Some clustering algorithms will perform both tasks. During implementation, the vital difference between standard clustering approaches and transformation based classifiers is the objective function. In the standard clustering approach, the objective function is minimized for reducing the intra-class distance. In the transformation based classification technique, Euclidean distance between data points and target is assumed as fitness function. Based on the fitness, each data point is iteratively updated. The classification accuracy differs in each iteration and highest accuracy is attained at particular iteration; that iteration umber is considered as ideal value for MAXIT. This process is carried out for all the SI algorithms. For example, for CSA this process is represented in Fig. 4 and this process is the key for success behind the transformation technique based classifiers using SI algorithms. More interpretations can be made through the scatter plots given in Figs. 8 and 9.

Two dimensional data are easier to represent and examine. So only the normalized mean and variance are considered for plotting the scatter plot. For easy analysis, 50 HEM and 50 ALL microscopic images are considered and mean and variance are extracted from each image. The initial data points before applying the CSA based transform is represented in Fig. 8 and they are non-linearly separable. The same can

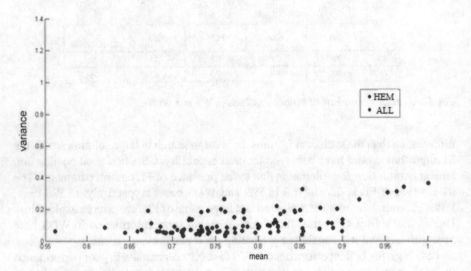

Fig. 8 Data points before applying CSA based transform

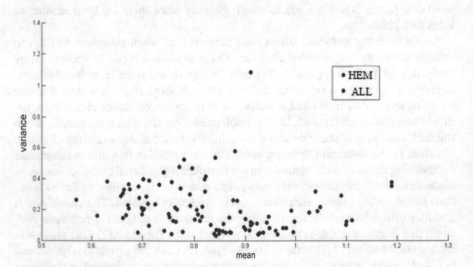

Fig. 9 Data points after applying CSA based transform

be witnessed through the scatter plot. These initial data points are updated iteratively using CSA algorithm, and stopped after seven iterations (ideal MAXIT = 7 in WS case). Now the data points are again plotted in scatter plot as represented in Fig. 9 and when compared to Fig. 8, this plot has data points which are more linearly separable i.e. separation between HEM and ALL data points can witnessed clearly. When the MAXIT is not equal to ideal MAXIT, then poor classification accuracy is observed.

6 Conclusion

This research work investigates the performance of CSA as transformation technique based leukemia classification. The results clearly show the significant performance of CSA as transformation technique based classifier with 87% of accuracy when compared to popular cluster techniques like K-Means and FCM. The performance metrics of each classifier is analyzed in two cases namely WS and WOS. The goodness of CSA as transformation technique over the CSA clustering technique is established. The transformation ability of CSA to convert non-linearly separable data points into linearly separable data points is proven using scatter plot. This study can be extended in the direction of building more precise classifier using hybridization and modification of original algorithms in appropriate manner.

References

1. Liu, L., Long, F.: Acute lymphoblastic leukemia cells image analysis with deep bagging ensemble learning. https://www.biorxiv.org/content/https://doi.org/10.1101/580852v1.full, https://doi.org/https://doi.org/10.1101/580852 (2019. Accessed on 17 Jan 2020
2. Pui, C.H.: Acute Lymphoblastic Leukemia, pp. 39–43. Springer Berlin Heidelberg, Berlin, Heidelberg (2017). https://doi.org/https://doi.org/10.1007/978-3-662-46875-3_57
3. Chatarwad, S., Bansode, P., Burade, A., Chaware, T.S.: Blood cancer detection using image processing. Int. J. Adv. Res. Electron. Commun. Eng. (IJARECE) 7(5) (2018)
4. Prellberg, J., Kramer, O.: Acute Lymphoblastic Leukemia Classification from Microscopic Images using Convolutional Neural Networks. ISBI 2019 C-NMC Challenge: Classification in Cancer Cell Imaging, Lecture Notes in Bioengineering book series (LNBE), pp. 53–61 (2019)
5. Fal Desai, P.G., Shet, G.: Detection of leukemia using image processing. Int. J. Adv. Res. Sci. Eng. 07(03) (2018)
6. Joshi, M.D., Karode, T.S., Suralkar, S.R.: White blood cells segmentation and classification to detect acute Leukemia. Int. J. Emerg. Trends Technol. Comput. Sci 2(3)
7. Kennedy, J., Eberhart, R · Swarm Intelligence. Morgan Kaufmann Publishers, San Francisco (2001)
8. Tharwat, A.: Classification assessment methods. Applied Computing and Informatics (2018). https://doi.org/10.1016/j.aci.2018.08.003
9. Oliva, D., Hinojosa, S., Cuevas, E., Pajares, G., Avalos, O., Gálvez, J.: Cross entropy based thresholding for magnetic resonance brain images using Crow Search Algorithm. Expert Syst. Appl. 79, 164–180 (2017)
10. Liu, D., Liu, C., Fu, Q., Li, T., Imran, K., Cui, S., Abrar, F.: ELM evaluation model of regional groundwater quality based on the crow search algorithm. Ecol. Indic. 81, 302–314 (2017)

11. Askarzadeh, A.: A novel metaheuristic method for solving constrained engineering optimization problems: Crow search algorithm. Comput. Struct. **169**, 1–12 (2016)
12. Parker, J.K., Hall, L.O.: Accelerating Fuzzy-C means using an estimated subsample size. IEEE Trans. Fuzzy Syst. **22**(5) (2014)
13. Taufik, A., Syed Ahmad, S.S.: A comparative study of Fuzzy C-Means And K-Means clustering techniques. In: 8th MUCET, Melaka, Malaysia (2014)
14. Fielding, A.H.: Cluster and Classification Techniques for the Biosciences. Cambridge University Press, Cambridge (2006)
15. Balavand, A., Kashan, A.H., Saghaei, A.: Automatic clustering based on Crow Search Algorithm-Kmeans (CSA-Kmeans) and Data Envelopment Analysis (DEA). Int. J. Comput. Intell. Syst. **11**, 1322-1337 (2018)

Development of Optical Parametric Oscillator Source for Investigating Two-Photon Excitation PDT

Masaki Yoshida, Yuichi Miyamoto, and Masahiro Toida

1 Introduction

1.1 Absorption Spectrum Characteristics

PDT (Photodynamic therapy) is a treatment for selectivenecrosis only tumors to generate active oxygen such as singlet oxygen by irradiating a laser of the absorption wavelength of the Q band. it causes a photodynamic reaction intravenously administered a photosensitive substance with high tumor affinity in vivo.

The photosensitive substance used in this time is Talaporfin sodium (Laserphyrin®), a second-generation photosensitive substance which generates active oxygen from photosensitive substances by irradiating light in the Q band. Laserphyrin® is excited at 405 nm, fluorescence of 672 nm peak occurs. However, the 405 nm light does not penetrate into tissues because the light is absorbed by hemoglobin and is strongly scattered in tissues. Therefore, the 667 nm light at Q-band is used in PDT. These absorption wavelengths are shown in Fig. 1.

1.2 Excitation Mechanism

Photosensitive material irradiated with laser light absorbs the energy, transitions from the ground state to the excited singlet state. Since the excited singlet state

M. Yoshida (✉)
Division of Medical Science, Graduate School of Medicine, Saitama Medical University, Saitama 350-1241, Japan
e-mail: pink-spider1173@outlook.jp

Y. Miyamoto · M. Toida
Department of Biomedical Engineering, Faculty of Health and Medical Care, Saitama Medical University, Saitama 350-1241, Japan

© Springer Nature Switzerland AG 2021
C. T. Lim et al. (eds.), *17th International Conference on Biomedical Engineering*,
IFMBE Proceedings 79, https://doi.org/10.1007/978-3-030-62045-5_15

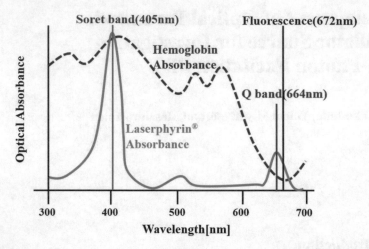

Fig. 1 Laserphyrin® and Hemoglobin absorptive properties and Laserphyrin® fluorescence

is very unstable, some release energy as fluorescence, returns to the ground state. This phenomenon is used as a photodynamic diagnosis (PDD) because there is a photosensitive substance in the tumor. It also transitions to a triplet state by the intersection between the terms of energy. Part of photosensitive substances in the triplet state release phosphorescence and return to the ground state, but on the other hand energy transitions to the triplet oxygen present in the tissue. The triplet oxygen subjected to energy transition is excited and changes to singlet oxygen (Fig. 2). Cancer tissue is oxidized by this singlet oxygen.

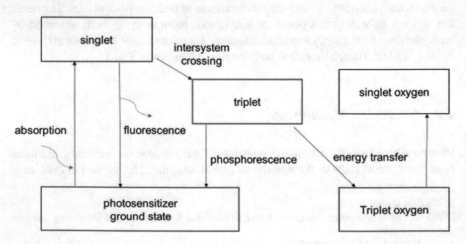

Fig. 2 Energy reaction process of photosensitizer

1.3 Two-Photon Excitation

Two-photon excitation is a phenomenon in which two photons are simultaneously absorbed by a molecule (two-photon absorption) to cause excitation. It was theoretically proposed by Maria Goppert-Mayer in the 1930, but at the time it was considered a phenomenon that could not exist. It was first observed experimentally in 1961 [1]. Since the transition rate of two-photon excitation is proportional to the square of the intensity of the excitation light, molecules can be spatially selectively excited at the micrometer level by using a focused laser beam. Further, it is possible to excite with near-infrared light because two photons with half the transition energy are used. Since long wavelength light having high permeability to a living tissue can be used, improvement in the tissue depth of the light action can be expected. It also affects normal cells in the optical path other than the lessoned tissue. Two-photon excitation PDT is being studied as a new treatment to solve these problems [2–4].

2 Purpose

PDT is only adapted an early superficial cancer, a tumor invading deeply in tissues is not applicable. The peak absorptive wavelength of Laserphyrin® is 405 nm, but this light can not be used because of low penetration in tissues.

In this study, we focused on these points and development of Optical Parametric Oscillator Source for investigating two-photon excitation PDT using near infrared light excitation.

3 Methods

3.1 Structure of KTP-OPO

Nd: YAG SHG-excited KTP-OPO was configured as the two-photon excitation light source. (Fig. 3) In this study, a KTP crystal was employed from various nonlinear optical crystals. In optical parametric interaction, the wavelengths of Signal light and Idler light depend on the refractive index. There are methods for changing the refractive index include angle tuning and temperature tuning. Temperature tuning lacks wavelength setting accuracy because there is an error between the set temperature and the actual crystal temperature. However the KTP crystal has a refractive index almost independent of temperature and the wavelength can be controlled only by the angle. We can select Signal light or Idler light by replacing the output coupler (OC), turn the wavelength by rotating the KTP.

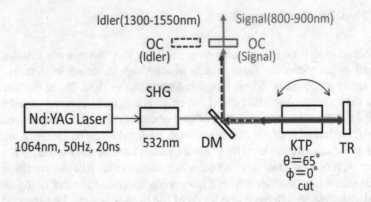

Idler(1300-1550nm)   Signal(800-900nm)

OC   OC
(Idler)   (Signal)

SHG

Nd:YAG Laser

1064nm, 50Hz, 20ns   532nm   DM   KTP   TR
                                    θ=65°
                                    φ=0°
                                    cut

Fig. 3 KTP OPO brock diagram

3.2 Two-Photon Excitation Coaxial Fluorescence Measurement System

The excitation-detection measurement optical system is shown in Fig. 4. If the excitation wavelengths are different, the focus position shifts in the optical axis direction. In orthogonal measurement, the adjustment of the focus position of the fluorescent light to the PD (Photodetector) in the direction orthogonal to the optical axis becomes complicated, but in coaxial measurement, the shift in the optical axis direction makes it easier to match the light receiving diameter of the PD with the focusing diameter. In this experiment, a coaxial rather than orthogonal fluorescence measurement system was employed.

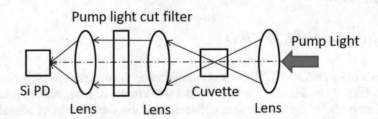

Pump light cut filter

Pump Light

Si PD
Lens   Lens   Cuvette   Lens

Fig. 4 Coaxial fluorescence detection block diagram

4 Results

4.1 KTP-OPO Oscillation Output Property

Figure 5 shows the oscillation output property of 810 nm light and 1328 nm light with respect to the pump light output. Figure 6 shows the spectrum width of 810 nm light and 1328 nm light.

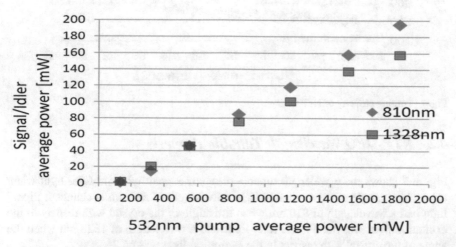

Fig. 5 Output property of Signal: 810 nm and Idler:1328 nm

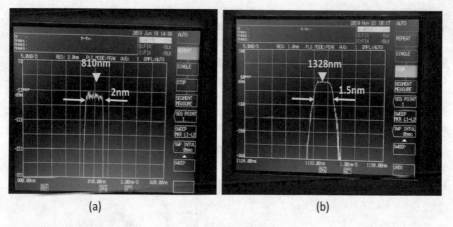

Fig. 6 Spectra of Signal: 810 nm (**a**) and Idler: 1328 nm (**b**)

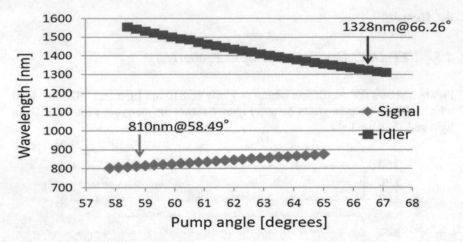

Fig. 7 Turning property of KTP OPO

4.2 KTP-OPO Wavelength Tunable Property

Figure 7 shows the wavelength tunable range of Signal light and Idler light when the incident angle of the excitation light wavelength of 532 nm is changed. Signal light had a wavelength of 810 nm when the angle of the crystal with respect to the excitation light was 58.49° and Idler light had a wavelength of 1328 nm when the angle of the crystal with respect to the excitation light was 66.26°.

4.3 Excited Light Focusing Spot Diameter

In the experimental system shown in Fig. 8, the convergent spot diameter of KTP-OPO at 810 nm was measured with a convex lens of f = 50 mm used in the two-photon excitation fluorescence measurement system. The diameter of the full-width half-maximum (FWHM) that is half of the maximum intensity is defined as the beam diameter. The FWHM in the x and y axes was about 70 μm. (Fig. 9).

Fig. 8 Spot size
measurement block diagram

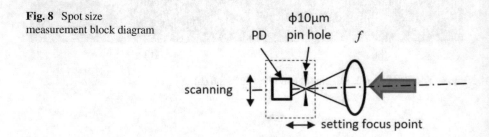

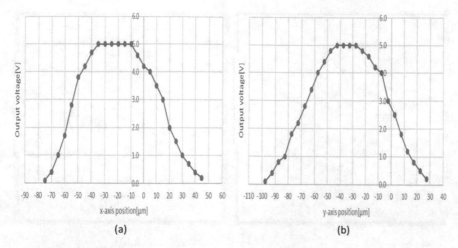

Fig. 9 Beam profile of x-axis position (**a**) and y-axis position (**b**)

5 Conclusion

In this study, we developed an optical parametric light source Nd: YAG SHG pumped KTP-OPO for two-photon pumped PDT search. The maximum average power of 810 nm and 1328 nm light was 200 mW. Since the pulse width is 20 ns and 50 Hz, the pulse power is 200 mW/50 Hz = 4 mJ/p and the peak power is 4 mJ/20 ns = 200 kW. Signal light and Idler light can be selected by exchanging the output mirror (OC), and the wavelength can be selected by rotating the KTP. Signal light had a wavelength of 810 nm when the angle of the crystal with respect to the excitation light was 58.49° and Idler light had a wavelength of 1328 nm when the angle of the crystal with respect to the excitation light was 66.26°.

In the future, we will proceed with fluorescence observation with Laserphyrin® aqueous solution and verification of cancer cell in vitro two-photon PDT and in vivo two-photon PDT.

References

1. Göppert-Mayer, M.: Ann. Phys. **401**, 273–294 (1931)
2. Konig, K., et al · Proc. SPIE **3592**, 43 (1999)
3. Liu, J., et al.: J. Photochem. Photobiol. B **68**, 156 (2002)
4. Karotki, et al.: Photochem. Photobiol. **82**,443 (2006)

Skeletal Bone Age Assessment in Radiographs Based on Convolutional Neural Networks

Jiaqing Wang, Liye Mei, and Junhua Zhang

1 Introduction

Bone age assessment plays an important role in the growth and development of children and adolescents [1–3]. Currently, bone age assessment is primarily conducted by trained radiologists, who manually assess hand bones in X-ray images in accordance with Greulich and Pyle (G&P) method [4] or Tanner-Whitehouse (TW) method [5]. The G&P method assesses the age by comparing the hand X-rays with the atlas consisting of reference images from subjects of different ages. Figure 1 shows the regions of interest (ROIs) of the G&P method. The TW method considers a specific set of ROIs which shows in the Fig. 2, the development of each ROI is divided into different stages and each stage is given a letter corresponding to a numerical score that varies by race and gender. Overall skeletal maturity score could be calculated by adding the scores for all ROIs.

Both methods are time consuming—an experienced radiologist may spend 1.4 mins on average to assess a patient with the G&P method and 7.9 mins for the TW2 [6]. Furthermore, both methods suffer from high intra- and inter-observer variability. The average spreads of the reading are 0.96 years (11.5 months) for the G&P method and 0.74 years (8.9 months) for the TW2 method [7].

Most of the existing automated bone age detection methods are based on the ROI of the clinical TW method for automatic feature extraction of the hand bone X-ray image. At the beginning of the twentieth century, Pietka et al. [8–9] proposed an epiphycis/metaphysis ROI (EMROI) and carpal ROI (CROI) segmentation method based on self-organized phalanx distance extraction, and used a fuzzy classifier to perform TW-level assignments on 360 non-public X-rays of 0 to 6-year-old children's

J. Wang · L. Mei · J. Zhang (✉)
School of Information Science and Engineering, Yunnan University, Yunnan, Kunming 650500, China

J. Wang
e-mail: jiaqing_wang@outlook.com

© Springer Nature Switzerland AG 2021
C. T. Lim et al. (eds.), *17th International Conference on Biomedical Engineering*,
IFMBE Proceedings 79, https://doi.org/10.1007/978-3-030-62045-5_16

Fig. 1 ROIs of the G&P
method

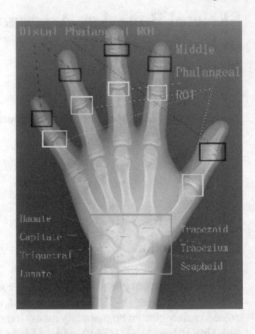

Fig. 2 ROIs of the TW
method

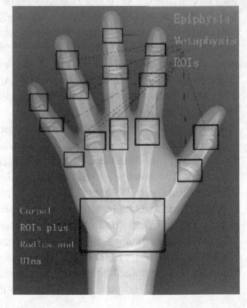

hand bones. The Mean Absolute Errors (MAEs) of the radiographic images were
2.41 years and 1.93 years. Gertych et al. [10] proposed a fuzzy logic processing
method in 2007, which was tested on a public dataset of 1,400 hand bone X-rays of
children aged 0 to 18 years, with a MAE of 2.15 years. Seok et al. [11] proposed
a method of decision rules in 2016. The mean square error (MSE) tested on 135

unpublished X-ray non-public data sets of specific age was 0.19 years. In addition, BoneXpert [12] proposed by Thodberg combined with the unified model of the TW and the G&P methods for automatic age assessment. It was tested on a non-public dataset of hand bone X-rays of 1559 children aged 7 to 17, with MSE of 0.42 years (G&P) and 0.80 years old (TW2). In 2017, Spampinato et al. [13] proposed a deep learning bone age detection method which called Bonet to perform automated bone age assessment on a public data set covering all age ranges, races and genders from 0 to 18 years old. The result shows the MAE is 0.8 years.

Most of them assess bone age by extracting features from the bones (either EMROIs or CROIs or both of them) commonly adopted by the TW or G&P clinical methods, thus constraining low-level (i.e., machine learning and computer vision) methods to use high-level (i.e., coming directly from human knowledge) visual descriptors. This semantic gap usually limits the generalization capabilities of the devised solutions, in particular when the visual descriptors are complex to extract as in the case of mature bones. There are also experimental data sets in the evaluation work that are not public and the amount of data is small, which is targeted to a specific gender or species, and most of them are subjects of a smaller age range whose bones have not yet been fused, among which the BoneXpert method also requires image quality. The above factors cause the experimental results to not truly reflect the performance of the automated detection method and the detection method is not beneficial Verification and promotion.

Aiming at the limitations of the above-mentioned clinical and automated bone age detection methods, this paper uses an improved end-to-end deep learning model Inception Resnet v2 to avoid the complexity and limitations of artificial feature extraction, without the need for any prior information. The X-ray image is used as the input of the neural network after simple preprocessing. The optimized convolutional neural network automatically extracts the deep features of the hand bone X-ray picture for recognition and classification processing. It is trained and tested on the public data set to achieve automatic bone age detection.

2 Method

2.1 Overview

We proposed an improved Inception Resnet v2 network to complete the bone age assessment [14]. The pipeline of our method is shown in Fig. 3. We deleted the softmax layer to optimize the network structure, took advantage of the excellent performance of the Inception-Resnet module in processing images, and introduced the CBAM based on the original network. The channel attention module and the space attention module were connected in series to improve the accuracy in bone age assessment.

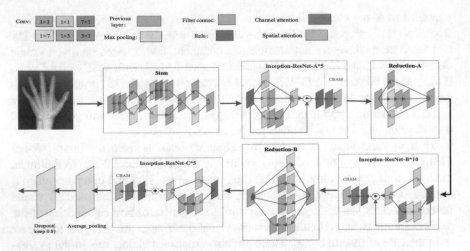

Fig. 3 Optimized Inception Resnet v2 network structure

2.2 The Improved Inception-Resnet Module

The core of the Inception module is to replace the original full connection mode by using the local optimal sparse structure and replace the traditional convolution layer with a large number of filters with a multilayer perceptron, avoiding redundancy to the maximum extent [15]. Compared with the single layer convolution model, the multilayer perceptron can make more cumbersome computation for each local sensing domain, thus approaching the desired result. It uses 1×1, 3×3, 5×5 convolution kernel and 3×3 Maximum pooling operation [16]. The results of different scale convolution kernel are jointly filtered with the pooling results, and the output is obtained. However, as the 5×5 convolution kernel needs a huge amount of calculation, all the convolution kernels are replaced by 3×3 [17]. After that, a series of smaller-scale convolution kernels are proposed. The introduction of the Inception module increases the width and depth of the network, but also causes problems such as gradient diffusion or gradient explosion. Here, the Inception Resnet v2 network solves the side effects of increasing the depth of the Convolutional Neural Network (CNN) by introducing residual connections to the improved Inception module. The optimized Inception-Resnet module is shown in Fig. 4.

2.3 The Application of CBAM

In order to improve the feature extraction capability of CNN models, Sanghyun Woo et al. Proposed a CBAM, which applies attention to both channel and spatial dimensions in 2018 [18]. In this paper, the CBAM is embedded into the Inception Resnet v2 network structure without significantly increasing the amount of calculations and

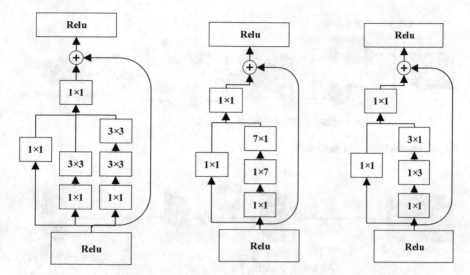

Fig. 4 Inception-Resnet modul

parameters. The combination of the Inception-Resnet module and the feature extraction capability of bone age recognition tasks is improved. The structure is shown in Fig. 5.

The significance of the channel attention module is to increase the weight of effective channels and reduce the weight of invalid channels. It performs the global pooling operation on the channel dimension and then obtains the weight through the same multi-layer perceptron and adds it as the final attention vector. Based on the average pooling, the maximum pooling is added, and two one-dimensional vectors can be obtained after two pooling functions. Global average pooling has feedback for each pixel on the feature map, and global max pooling performs gradient backpropagation calculations. Only the feature map with the largest response in the feature

Fig. 5 Inception-Resnet module and improved CBAM-Inception-Resnet module

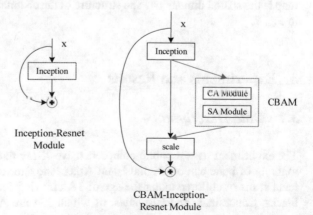

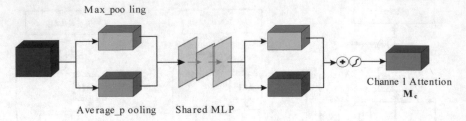

Fig. 6 Channel attention module

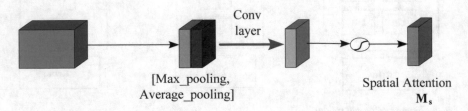

Fig. 7 Spital attention module

map has gradient feedback, which can be used as a supplement to global average pooling. The structure of the channel attention module is shown in Fig. 6.

In addition to generating the attention model on the channel, the CBAM also requires that the network be aware of the response in the feature map at the spatial level. The spatial attention module uses average pooling and max pooling to perform channel-level compression operations on the input feature map, and performs mean and max operations on the channel features in the channel dimension, respectively. Two two-dimensional features were obtained, which were stitched together according to the channel dimensions to obtain a feature map with two channels, and then a convolutional operation was performed on the hidden layer containing a single convolution kernel to ensure the final feature. It is consistent with the input feature map in the spatial dimension. The structure of the channel attention module is shown in Fig. 7.

3   Experiments and Results

3.1   Data Description

The experiment uses a public comprehensive X-ray data set for automatic benchmarking of bone age on Digital Hand Atlas. The dataset contains 1391 X-ray left-hand scans of children under the age of 18, covering four ethnic groups of Asians, blacks, Caucasians, and Hispanics, of which 334 are Asians, 359 are blacks, 333 are Caucasians, and 365 are Hispanics, 700 men and 691 women. Each X-ray scan image is provided by an expert radiologist with a bone age value.

This data set is a public data set, which is convenient for comparison of network performance. The data set is highly applicable and covers a wide range of ethnic groups and a wide range of ages. Therefore, the network trained by this data set can be applied to bone age detection of various age groups and races, and can make up for the shortcomings of current automatic bone age assessment methods to a certain extent.

3.2 Data Preprocessing

In this paper, because the original appearance of the ROI features of the hand bone X-ray image can not be changed, data augmentation methods such as random stretching are not selected. Instead, the image is flipped up and down by 180 degrees to expand the experimental data volume double.

The Z-Score standardized processing is performed on the X-ray hand bone pictures after data enhancement. The processed data has a mean value of 0 and a standard deviation of 1. The data of different magnitudes are transformed into a unified metric, which improves data comparability and weakens the data Explanatory.

3.3 Selection of Loss Function

The experiment uses a fixed learning rate to train the model. Considering the model convergence at a fixed learning rate, the experiment uses the MSE loss function. The gradient of the MSE loss will decrease as the loss decreases, and it is not easy to miss at the minimum point when the gradient decline is about to end. The training using the MSE loss function is more accurate than other loss functions.

The detection loss is defined as the MSE between the bone age reading of the hand bone x-ray picture detection and the true value of the label, and its formula is as follows:

$$MSE = \frac{1}{N} \sum_{i}^{N} \left(y_{true} - y_{pred} \right)^2 \tag{1}$$

N are the number of samples, y_{true} Represents the true value of the bone age of the label, y_{pred} Indicates the predicted bone age value. The larger the MSE loss function value, the worse the prediction result.

3.4 Improved Cross-Validation Method

The data set used in the experiments covers four ethnic groups of Asians, blacks, Caucasians, and Hispanics, and the characteristic regions of hand X-ray pictures of different races are different. Ordinary K-fold cross validation (see Fig. 8) method divides the data set into K equal parts in order. For each time (a total of k times), there will be (K-1) copies for the training set, and the remaining one for the validation set, and then K models will have K results, the average of the K results is the final result. This method divides the data set in equal order, and each fold may have an imbalanced category allocation problem, and the verification effect is meaningless. Based on this, the experiment uses a stratified K-fold cross validation method (as shown in Fig. 9), hierarchically sampling the data to ensure that the proportion of samples in each category in the training set and test set is the same as the original data. The improved cross validation makes the experimental results more accurate.

Fig. 8 K-folds cross validation

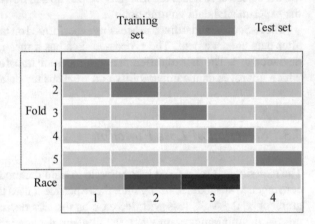

Fig. 9 Stratified K-folds cross validation

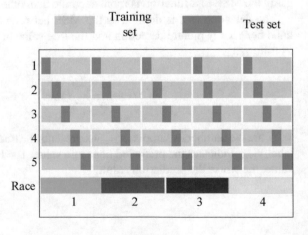

3.5 Metric

For the bone age numerical regression prediction results, the MAE index was used to evaluate the experiment, and its formula is as follows:

$$MAE = \frac{1}{N}\left(\sum_{i=1}^{N}|y_{true} - x_{pred}|\right) \tag{2}$$

N are the number of samples, y_{true} represents the true value of the bone age of the label, y_{pred} represents the predicted bone age value. The smaller the value of MAE, the better the prediction model has.

3.6 Experimental Platform and Training Strategies

The model in this paper is performed on a Windows 10 computing platform, It includes the Intel (R) Core (TM) i7-8700 K model CPU, the 16G × 2 memory and the NVIDIA GeForce GTX 1080TiGPU, all programs are open source framework keras with tensorflow as its backend and its python language interface implementation.

The model is trained using the Adaptive Moment Estimation algorithm under the MSE loss function. The batch size is set to 32, the initial learning rate is 0.1, and the learning rate is gradually reduced at a rate of 5 or 10 times each time to adjust the learning rate. When the better results appear, fine-tune the learning rate to obtain the best possible effect. Use the bone age label value as the target of model training. Use the stratified five-folds cross validation method to evaluate the performance of the network model on the enhanced data set. Perform 200 iterations for a single fold, and the model takes MAE as output.

3.7 Experimental Results and Analysis

The learning rate controls the speed of adjusting the weight of the neural network based on the loss gradient. The learning rate is too large, and the magnitude of the gradient descent easily crosses the optimal solution. The learning rate decreases, the speed of the gradient descent slows down, and the convergence time increases.

Aiming at the application of the improved Inception Resnet v2 model in bone age detection, in order to find the optimal learning rate as much as possible, a large amount of parameter adjustment work was performed experimentally. The test results obtained by the parameter adjustment are shown in Fig. 10.

In order to test the performance of this method, a comparison is made with the deep learning method Bonet [13] applied on the same data set. In 2017, Spampinato et al. [13] proposed a bone age detection method Bonet based on deep learning. Bonet

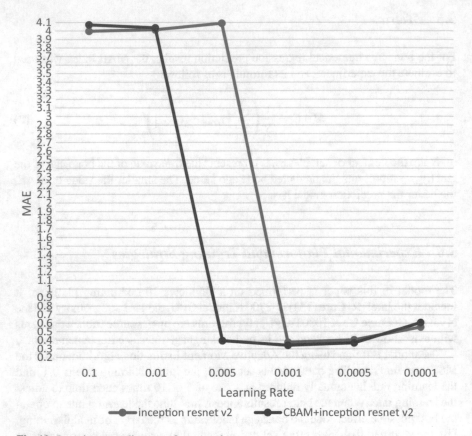

Fig. 10 Learning rate adjustment for network

consists of 5 convolutional layers and a deformed layer after the fourth convolutional layer. The convolutional network consists of a fully connected layer of 2048 neurons and a single neuron that provides an estimate. The regression network uses ordinary K-folds crossing validation. This method is the first to carry out automated bone age assessments tested on a public data set and across all age ranges, races and genders, and represents a baseline in the field. The experimental results compared with this method are shown in Table 1.

In the case of using a more advanced stratified K-folds cross validation method, the improved Inception Resnet v2 method in this paper has doubled the accuracy of

Table 1 Model performance comparison

Mode	Mae
Inception Resnet v2	0.377
CBAM + Inception Resnet v2	0.3416
Bonet	0.8

bone age detection achieved by the Bonet network. The same applies to public data sets covering all races, ages and genders. The method in this paper can extract more distinguishing features, so the recognition rate is higher, and it has better robustness and generalization.

The good results of the Inception Resnet v2 network in bone age detection prove that the Inception-Resnet module is superior in image processing. At the same time, the introduction of the CBAM module has improved the performance of the original network in bone age detection to a certain extent.

4   Conclusion and Future Work

In this paper, deep learning method is used to realize bone age recognition of hand bone X-ray pictures. By optimizing the deep CNN model to reduce parameters, the complexity and limitations of artificial feature extraction are avoided, and accuracy is improved. The data augmentation method used effectively avoids the overfitting problem that occurs in deep learning algorithms when the sample size is insufficient. The selected loss function and evaluation index apply the improved CNN to the bone age detection experiment, and it also uses a more feasible stratified five-folds cross validation method. The experiment proves that the method in this paper improves the recognition rate, has better robustness and generalization, and meets the higher clinical requirements for bone age detection to a certain extent. Considering the end-to-end method used in this paper, how to visualize the feature extraction area of the CNN and compare with the clinical method will be the next research direction.

References

1. Martin, D.D., Deusch, D., Schweizer, R., et al.: Clinical application of automated Greulich-Pyle bone age determination in children with short stature[J]. Pediatr. Radiol. **39**(6), 598–607 (2009)
2. Liu, J., Liu, J.: A review of the development of bone age assessment system[J]. Life Science Instruments **15**(2), 9–13 (2017)
3. Zhang, A., Gertych, A., Liu, B.J.: Automatic bone age assessment for young children from newborn to 7-year-old using carpal bones[J]. Comput. Med. Imaging Graph. **31**(4–5), 299–310 (2007)
4. Garn, S.M.: Radiographic atlas of skeletal development of the hand and wrist[J]. Am. J. Hum. Genet. **11**(3), 282 (1959)
5. Kim, S.Y., Oh, Y.J., Shin, J.Y., et al.: Comparison of the Greulich-Pyle and Tanner White-house (TW3) Methods in Bone Age Assessment[J]. Journal of Korean Society of Pediatric Endocrinology **13**(1), 50–55 (2008)
6. Christoforidis, A., Badouraki, M., Katzos, G., et al.: Bone age estimation and prediction of final height in patients with β-thalassaemia major: a comparison between the two most common methods[J]. Pediatric Radiology. **37**(12) (2007)

7. King, D. G., Steventon, D. M., O'sullivan, M.P. et al.: Reproducibility of bone ages when performed by radiology registrars: an audit of Tanner and Whitehouse II versus Greulich and Pyle methods[J]. British J. Radiol. **67**(801), 848–851.

8. Pietka, E., Gertych, A., Pospiech, S., et al.: Computer-assisted bone age assessment: Image preprocessing and epiphyseal/metaphyseal ROI extraction[J]. IEEE Trans. Med. Imaging **20**(8), 715–729 (2001)

9. Pietka, E., Pospiech-Kurkowska, S., Gertych, A., et al.: Integration of computer assisted bone age assessment with clinical PACS[J]. Comput. Med. Imaging Graph. **27**(2–3), 217–228 (2003)

10. Gertych, A., Zhang, A., Sayre, J., et al.: Bone age assessment of children using a digital hand atlas[J]. Comput. Med. Imaging Graph. **31**(4–5), 322–331 (2007)

11. Seok, J., Kasa-Vubu, J., DiPietro, M., et al.: Expert system for automated bone age determination[J]. Expert Syst. Appl. **50**, 75–88 (2016)

12. Thodberg, H.H., Kreiborg, S., Juul, A., et al.: The BoneXpert method for automated determination of skeletal maturity[J]. IEEE Trans. Med. Imaging **28**(1), 52–66 (2008)

13. Spampinato, C., Palazzo, S., Giordano, D., et al.: Deep learning for automated skeletal bone age assessment in X-ray images[J]. Med. Image Anal. **36**, 41–51 (2017)

14. Szegedy, C., Ioffe, S., Vanhoucke, V., et al.: Inception-ResNet and the impact of residual connections on learning[J]. arXiv preprint arXiv:1602.07261.

15. Szegedy, C., Liu, W., Jia, Y., et al.: Going deeper with convolutions[C]. In: Proceedings of the IEEE conference on computer vision and pattern recognition, pp 1–9 (2015)

16. Ioffe, S., Szegedy, C.: Batch normalization: Accelerating deep network training by reducing internal covariate shift[J]. arXiv preprint arXiv:1502.03167 (2015)

17. Szegedy, C., Vanhoucke, V., Ioffe, S., et al.: Rethinking the inception architecture for computer vision[C]. In: Proceedings of the IEEE conference on computer vision and pattern recognition, pp. 2818–2826 (2016)

18. Woo, S, Park, J., Lee, J.Y., et al.: Cbam: Convolutional block attention module[C]. In: Proceedings of the European Conference on Computer Vision (ECCV), pp. 3–19 (2018)

An Evaluation on Effectiveness of Deep Learning in Detecting Small Object Within a Large Image

Nazirah Hassan, Kong Wai Ming, and Choo Keng Wah

1 Introduction

Artificial Intelligence (AI) that began with neural networks existed many years ago. The success of AI increased with the invention of machine learning (ML) which is a method of data analysis that automates analytical model building. ML is a branch within AI and it was invented with the idea that systems can learn from data, detect objects, identify patterns and make decisions with minimal human intervention. However, recently, the success of AI has been booming due to an invention called Deep Learning (DL). Deep Learning is a subset within Machine Learning. It consists of networks that are capable of learning from unsupervised data that is unstructured or unlabeled. The development of Deep Learning has been accelerating over the past few years and many complex tasks can now be achieved using Deep Learning. Examples of these complex tasks are recognition and detection of objects which are now achievable to a large degree due to the revolution of deep learning: neural networks that rely on automatic feature extraction through convolutional layers. Deep Learning can be used within autonomous vehicles [1, 2] and for classifying different types of animals [3, 4] and diagnosing stages of cancer [5].

However, despite the high success rate of the above-mentioned applications as well as the progress of the deep learning field, the detection of small objects within a large-sized image ranging from 4,096 × 3,072 pixels to 7,680 × 4,320 pixels in ultra-high-definition frame remains a challenge. One such example is the accurate detection and classification of pedestrian traffic light (PTL) signals. A typical 2D

N. Hassan (✉) · K. W. Ming · C. K. Wah
School of Engineering, Nanyang Polytechnic, Nanyang, Singapore
e-mail: nazirah_hassan@nyp.edu.sg

K. W. Ming
e-mail: kong_wai_ming@nyp.edu.sg

C. K. Wah
e-mail: choo_keng_wah@nyp.edu.sg

© Springer Nature Switzerland AG 2021
C. T. Lim et al. (eds.), *17th International Conference on Biomedical Engineering*,
IFMBE Proceedings 79, https://doi.org/10.1007/978-3-030-62045-5_17

175

colored image consists of three channels—red, green and blue (RGB) and it is made up of pixels that range from 0 to 255 within each channel. These pixels make up the input layer of deep learning neural networks where with each progressing layer, it goes through feature extractions and non-linearity functions. The input images would be resized to a smaller dimension such as $1,024 \times 1,024$ pixels [Faster R-CNN] or 300×300 pixels [SSD models]. In the case of pedestrian traffic light signals which are relatively small objects, it may become unrecognizable if the source image is overly compressed, making the detection of PTL signal a near impossible task. In addition, the existence of objects similar to the shape or form and colors of PTL signals further complicate the detection process.

2 Related Works

The most commonly used approach in detecting pedestrian traffic light signals is the classical colour-based segmentation method such as those shared in [6, 7]. Reference [6] proposed a Traffic-Light Recognizer to support the visually impaired person using contour and colour-based approach in order to identify potential Active Output Unit candidatures. Reference [7] proposes a similar approach whereby the extraction algorithm is based on colour segmentation and the geometrical properties of the PTL signals. The recognition algorithm is then created based on SVM classification. These approaches yield fantastic results. However, different designs and forms of pedestrian traffic lights adopted in different countries may have impact on the robustness and accuracy of the detectors, making the deployment of such systems a challenge. Traffic lights that are partially occluded could also contribute to confusion in the detection process.

Currently there are machine learning and deep learning approaches deployed, such as the HSV-based analytic image processing and learning-based processing by [8] and Faster R-CNN-like models by [9]. Reference [9] addressed the challenge using a novel attention model based on a Faster R-CNN algorithm. The locator and recognizer then use another Faster R-CNN-like model. This has motivated us to explore the state-of-the-art deep learning approaches to provide a more robust solution that can be generalized for wider applications.

3 Proposed Concepts for Pedestrian Traffic Light Detection

In this paper, we propose a novel method in addressing the problem of detecting pedestrian traffic light signals This is done with the aim of addressing the issues raised in previous sections and in making the detection more efficient and robust, as summarized in Fig. 1.

We suggest two different methods of segmentation with the objective of identifying regions of interest (ROI) of traffic light from a large-sized image. The first

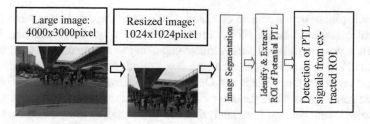

Fig. 1 Proposed concept of PTL signal detection approach

method consists of two stages. The first stage is to perform a segmentation onto the original image based on a classical Hue, Saturation and Value (HSV) threshold. Upon segmentation, the image then undergoes object detection using a R-CNN Object Detector. The second method, however, is a single stage method that consists of a Mask R-CNN Object Instance Segmentation module.

3.1 HSV- Based Image Segmentation with R-CNN Object Detection

Segmentation based on HSV. The HSV colour model was designed by computer graphics researchers in 1970s. It was designed in order to ensure that a colour representation model closely aligns with the way human vision perceives colour-making attributes. It consists of a cylindrical colour model that remaps primary colours into dimensions that are easier for humans to understand. These dimensions are termed as hue, saturation and value. Hue represents the different type of colours, the saturation represents the intensity of the colours while value represents the brightness of each colour.

Implementation. Pedestrian traffic light signals consist of a red man to show that pedestrians are not allowed to cross the road while a green man is used to show that it is safe to cross the road. Colours within captured images are expressed by the RGB colour model [10]. The RGB colour model shows colours in three different arrays: Red, green and blue arrays. Within each array, the value of the pixels ranges from 0 to 255. Pixels containing the value of 0 is black in colour while pixels containing the value of 255 is white in colour. Despite the common use of RGB colour model, viewing and analysing colours in the RGB colour model is not preferred. In order to view, analyse and segment colours the way humans perceive colours, the HSV colour model is preferred. Hence, the RGB values of the pedestrian traffic light signals are converted to HSV values via the formulae given in Ref. [11].

Using three different cameras with different camera specifications, 4000 images were captured from various locations and in different lighting conditions for the purpose of HSV threshold selection. The cameras used are an 8 Megapixels (MP) Raspberry Pi Camera, HTC 10 mobile phone with a camera of 9.1 MP and a Samsung

S9 mobile phone with a camera of 12MP. The images captured were taken in three different conditions: sunny skies, cloudy skies and when it is drizzling. Various perimeters were considered during image capturing to ensure careful consideration for the HSV thresholds. The night scene was excluded in this study as it involves a different set of imaging techniques. Out of the 4000 images captured, 2000 consists of green man signals while the other 2000 consists of red man signals. The range of HSV values were decided from analyzing the 4000 training images and the selected HSV range is tabulated in table (see Table 1).

From the range of HSV values decided, the image is segmented: the entire image is converted to black pixels aside from the objects that consists of colors within the HSV range specified (see Fig. 2). The segmented image then undergoes binary conversion where it is then diluted and eroded for clarity. After which the image undergoes labelling ("*bwlabel*") [12] and region proposals. Through region proposals, the features of the object such as area, centroid and boundaries were obtained.

The centroid of each segmented object is located, and a boundary is created around the object upon HSV segmentation and labelling. The image within the boundary is then extracted into a new image. These images (see Fig. 3) are then sent for object detection using the R-CNN object detection algorithm.

R-CNN Object Detection. R-CNN is a type of object detection network that is extended from a convolutional neural network (CNN). The evolution and highlight of deep learning are mainly due to the creation of CNNs. A CNN is a form of neural network that relies heavily on feature extraction layers called convolutional layers. Convolutional layers extract features from an image automatically based on filter size and number of filters that are determined by a neural net designer. The convolutional

Table 1 HSV range for red and green man PTL signals segmentation

Type	Range	Hue (°)	Saturation (%)	Value (%)
Red man	Min	329.76	70	55
	Max	36.00	100	100
Green man	Min	158.00	73.6	19.6
	Max	177.98	100	96.5

Fig. 2 An image after going through HSV segmentation

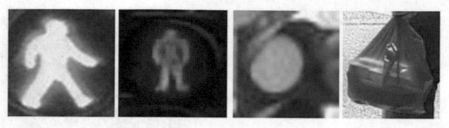

Fig. 3 New images created after HSV Segmentation, binary conversion, labelling and extraction for object detection using R-CNN

layers convolve the image input by sliding the convolutional filters along the input vertically and horizontally. As the convolution filter slides across the image input, the dot product of the weights and input (pixel values of image input) are computed, and a bias term is then added. As the layers get deeper, the network is capable of recognizing objects based on the features detected. Therefore, R-CNN is an extension of CNN as the foundation of the layers is still feature extraction layers. However, it is also incorporated with an algorithm to identify potential region of interests for automatic object localization.

R-CNN uses selective search and it is the simplest form of object detection method. Objects are searched only within 2000 regions using the Edge Boxes algorithm. For this research, the detector was trained using transfer learning in order to compensate for the insufficient number of training images and to ensure that the training can be executed successfully using a single graphic processing unit (GPU). The network was trained using transfer learning from AlexNet. AlexNet is the name of a CNN that was designed by Alex Krizhevsky [13]. It has a depth of eight layers. Despite the small number of layers, it has a parameter of 61 million. The image input size is 227 by 227 with a channel size of 3 (colored images). The detector was trained for 42 epochs on 4000 training images. The trained detector achieved a training accuracy of 100% with a loss of 1.4046×10^{-5}. After training, the detector is then tested against 400 test images. See Fig. 4 for an example of detection.

3.2 Mask R-CNN Object Instance Segmentation

Introduction to Mask R-CNN. Alternative solutions to the challenging task of object detection are the emerging deep learning algorithms. It ranges from fast and efficient algorithms such as Single Shot Detector (SSD) [14] and You Only Look Once (YOLO) [15], to the more advanced and accurate models of Region-Based CNN, such as Faster R-CNN [16].

The Mask R-CNN model was introduced by [17], evolving from the initial introduction of the Region-Based Convolutional Neural Network, or R-CNN. It supports both object detection and instance segmentation and is the most recent advancement of the R-CNN family models. The architecture of the Mask R-CNN (Fig. 5) consists

Fig. 4 Example of R-CNN
Detection Result

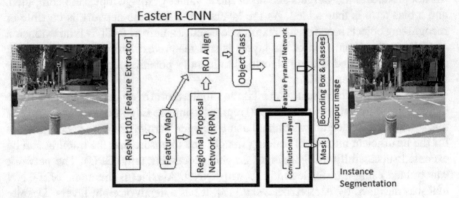

Fig. 5 Mask R-CNN architecture

of two key baseline systems, namely the Faster R-CNN module and the Instance Segmentation module. The Faster R-CNN module consists of a ResNet-101 [18] feature extraction module. The feature extraction module is a 101-layer deep convolutional neural network that is extended with an RPN (Regional Proposal Network). The ResNet-101 serves to extract the key features of object to be detected, pass them into the RPN to identify proposed regions where the object lies. The ROI Align module takes the object proposal and divides it into a certain number of bins where the data points are sampled and re-computed using bilinear interpolation. A Feature Pyramid Network (FPN) [19] is used to process the transformed features and to produce the final mask for the instance segmentation.

Implementation of Mask-RCNN. There are several implementations of the Mask R-CNN, one of which is a fully tested open-source implementation [20] and has been used successfully in many applications such as life sciences, automation and city imaging. The codes were adapted for the PTL signal detection and it was trained only for four classes as the detection pipeline was simplified to only look for traffic related signals. These include the following object classes: green car traffic lights, red

Fig. 6 Example of mask R-CNN object instance segmentation

car traffic lights, green man and red man. These object classes are selected to ensure that the network can clearly distinguish between the various traffic light signals available on the streets.

Based on the above implementation, the Mask R-CNN Segmentation network was tested on 400 test images. The results of the segmentation and detection of PTL signals are discussed in the following section. See Fig. 6 for an example of Mask R-CNN Object Instance Segmentation.

4   Results

Both methods were tested on the same test set which consists of four hundred test images; 200 images containing red man signals while the other 200 are green man signals. The test images were captured from various locations in Singapore to ensure a fair selection of test images. These images were also captured at different view distances of PTL. These distances range from one lane to seven car lanes, producing varying size of PTL in a typical street.

The results of the detection for both methods are tabulated below (see Table 2).

Based on the tabulated results above, the hybrid DL (Mask R-CNN) approach has a higher average accuracy of 99%. However, despite achieving a higher accuracy, the hybrid approach requires more computational resources. The HSV + R-CNN Detection method took 2 h for training while the Mask R-CNN Object Instance

Table 2 Detection results for both methods

Method	No. of PTL signals detected	Accuracy for green and red PTL signals detection	Averag accuracy
HSV + R-CNN Detection	Green: 197/200 Red: 185/200	Green: 98.5% Red: 92.5%	**95.5%**
Mask R-CNN Object Instance Segmentation	Green: 197/200 Red: 199/200	Green: 98.5% Red: 99.5%	**99%**

Segmentation method took 2.5 h for training. The time taken for training may differ by a small margin. However, the former method can be implemented on current devices such as Microsoft Surface Tablets that does not have a pre-installed GPU card. The same cannot be said for the hybrid approach as the method is too complex and it requires the use of GPU for execution.

Therefore, aside from computer desktops with pre-installed GPU cards, this method cannot be implemented easily especially in the use of PTL signal detection. PTL signal detection requires real-time implementation and easy execution in portable devices. Hence, the HSV-based approach is preferred.

5   Consideration for Real—Time Implementation

Based on the results obtained in the previous section, further research was carried out to ensure real-time implementation is possible in edge devices such as microcontrollers and mobile phones.

5.1   HSV-Based Segmentation New Model

The HSV-based Segmentation method equipped with a R-CNN Detector proves that the detection of PTL signals can be achieved without the need of a complex model. However, the question remains if a detector is necessary when location of the PTL signals within the image is not needed. Therefore, to ensure lesser computational cost and faster execution, the detection module is replaced with a CNN image classification module.

Image Classifier. An image classifier is a type of deep learning neural network (DNN) capable of classifying objects automatically without any prior image processing. This type of neural network has been greatly improved due to the breakthrough of CNNs. The difference between a detection network and a classification network is the absence of localization. In classification, the position of the object is not needed therefore, no algorithms are required to analyze the location or coordinates of each object. The primary focus of the network is to classify the objects into its category. Hence, for the purpose of PTL signal detection upon HSV Segmentation, an image classification network would be an ideal solution.

The image classifier is trained using the architecture of InceptionV3. The network was trained from scratch on seven object classes: green car traffic lights, red man, green man, red car traffic lights, LED Digits, background class for segmented green objects and background class for segmented red objects. The number of classes have expanded since classification networks do not have an automated background class. Therefore, the DNN designer needs to create strategic background classes to ensure that the network is accurate and precise enough to recognize the green and red man efficiently.

Table 3 Precision and recall results for image classification network

PTL Signal Color	Precision	Recall
Red	175/177 = 0.989	175/177 = 0.989
Green	196/196 = 1	196/196 = 1

Table 4 Detection results for HSV + image classifier

Methods	No. of PTL successfully segmented and classified	Accuracy for green and red PTL signals	Average accuracy (%)
HSV-based + CNN Classifier	Green Man: 196/200 Red Man: 175/200	Green Man: 98% Red Man: 87.5%	92.75%

Accuracy of Image Classifier. The image classification network was then tested against the same 400 test images. The accuracy, precision and recall of the network are tabulated below (see Tables 3 and 4).

Based on the tabulated results above, the accuracy of the image classifier is not as high as the previous implementation with a R-CNN Detector. The R-CNN Detector network has an automated background class that ensures high precision and recall due to the absence of False Positives and Negatives. However, despite achieving a higher accuracy, the implementation cannot be used for practical usage as it takes two seconds per detection unlike the DNN classification network that produces results instantaneously.

In order to achieve real-time implementation for a working pedestrian traffic light signal detector, the HSV + Image Classifier method is preferred. However, to compensate for the drop in accuracy, a minimum confidence of 90% is required to ensure the absence of False Positives and Negatives.

6 Real-Time Implementation of HSV + Image Classifier Model

Due to the rapid development of the deep learning field, there are many new products created to house deep learning networks for practical applications. Some examples of those devices are GPU-enabled mobile phones, Raspberry Pi and Jetson Boards. Jetson Boards are minicomputers that are designed and produced by NVDIA. For the purpose of a PTL signal detection, a portable device is preferred. Hence, the Jetson Nano module was chosen for this research.

6.1 Implementation on Jetson Nano

Jetson Nano is a minicomputer that has a quad-core processor for its CPU and a 128-core Maxwell for its GPU. The random-access memory of the device is four gigabytes and the device has a dimension of 69.6 mm by 45 mm. It is compact and based on the technical specifications stated, it has the capacity to run DNN models efficiently.

Firstly, the image classifier trained using Keras with a Tensorflow backend is converted to a frozen graph model (TensorRT format) to ensure fast and easy execution. TensorRT is an SDK for high-performance deep learning inference [21]. TensorRT optimizes neural network models such that it can perform up to 40 times faster than CPU-only platforms. The reason for the fast execution is the conversion of trained models from 32 floating points (FP32) to 16 floating points (FP16). Once the model has been converted, it is then used together with an HSV Segmentation script that gains continuous input from a webcam. The webcam is attached to the Jetson Nano via USB.

Using the Jetson Nano board and the HSV Segmentation + Image Classifier method, the PTL signal detection was made possible. The method achieved real-time implementation as it takes lesser than three milliseconds for a single detection. In order to ensure the absence of False Positives and Negatives, the DNN image classifier is equipped with thresholds. Once a detection is made, the probability or confidence of the detection results is analyzed. If the confidence of the detection is above 70%, the result is relayed to the user. However, if the confidence is below 70%, the information is not relayed to ensure user safety. Another image is then captured for processing. With this, the accuracy and precision of the detection program is sufficiently high such that no wrong detection can be made.

7 Conclusion

From this in-depth research, it can be seen that despite the breakthrough in deep learning, classical methods such as HSV segmentation (image processing) is still relevant. The accuracy and reliability of image processing methods can be improved using deep learning neural networks. In this case, the novel HSV segmentation process equipped with a DNN image classifier proves to be the most reliable solution for a real-time implementation of a pedestrian traffic light signal detector.

However emerging deep learning methods such as the Mask R-CNN does provide more contextual information as it can detect multiple objects without any colour restriction. Despite the advantages, current hardware limits real—time execution. This constraint, however, would no longer be present in the future due to advancement of technology that is coming up with more GPU enabled devices and the availability of 5G Network. We foresee a combination of these two innovations as a method to overcome current limitations that are mentioned above.

Based on the findings of this research and considering current limitations and opportunities for future development, we will continue to explore various algorithms and combination of methods to provide an accurate and efficient method for the detection of small objects within large size images. Going forward, our plan is to translate or quantize complex hybrid approaches in order to ensure that it can be used for real-time applications. Hybrid approaches provide contextual information that could be a key factor in making decisions for AI modules that are used in sensitive practical applications. However, for the detection of pedestrian traffic light signals, we are confident in the current implementation used within the Jetson Nano Board (HSV + DNN Image Classifier).

References

1. Al-Qizwini, M., Barjasteh, I., AlQassab, H., Radha, H.: Deep learning algorithm for autonomous driving using googlenet. In: Intelligent Vehicles Symposium (IV), 2017 IEEE, pp. 89–96. IEEE (2017)
2. Chen, C., Seff, A., Kornhauser, A., Xiao, J.: Deepdriving: Learning affordance for direct perception in autonomous driving. In: Proceedings of the IEEE International Conference on Computer Vision, pp. 2722–2730 (2015)
3. Chen, G., Han, T. X., He, Z., Kays, R., Forrester, T.: Deep convolutional neural network based species recognition for wild animal monitoring. In: IEEE International Conference on Image Processing (ICIP), pp. 858–862 (2014)
4. Gomez Villa, A., Salazar, A., Vargas, F.: Towards automatic wild animal monitoring: Identification of animal species in camera-trap images using very deep convolutional neural networks. Ecological Informatics **41**, 24–32 (2017)
5. Olliverre, N., Yang, G., Slabaugh, G., Reyes-Aldasoro, C.C., Alonso, E.: International Workshop on Simulation and Synthesis in Medical Imaging. Springer; Cham, Switzerland: 2018. Generating Magnetic Resonance Spectroscopy Imaging Data of Brain Tumours from Linear, Non-linear and Deep Learning Models, pp. 130–138 (2018)
6. Mascetti, S., Ahmetovic, D., Gerino, A., Bernareggi, C., Busso, M., Rizzi, A.: Robust traffic lights detection on mobile devices for pedestrians with visual impairment. Computer Vision Image Underst. https://doi.org/10.1016/j.cviu.2015.11.017 (2016)
7. Cheng, R., Wang, K., Yang, K., Long, N., Bai, J., Liu, D.: Real-time pedestrian crossing lights detection algorithm for the visually impaired. Multimedia Tools Appl.**77**(16), 20651–20671 (2018)
8. de Charette, R., Nashashibi, F.: Traffic light recognition using image processing compared to learning processes. In: Proceedings of the 22nd International Con- ference on Intelligent Robots and Systems, IEEE, pp. 333–338 (2009)
9. Lu, Y., Lu, J., Zhang, S., Hall, P.: Traffic signal detection and classification in street views using an attention model: Computational Visual Media, vol. 4, No. 3, pp. 253–266 (2018)
10. "Analog and Digital Images," Principles of Remote Sensing - Centre for Remote Imaging, Sensing and Processing, CRISP, 2001. [Online]. Available: https://crisp.nus.edu.sg/~research/tutorial/image.htm. Accessed 24 Sep 2019
11. "RGB to HSV conversion I color conversion", Rapidtables.com, 2019. [Online]. Available: https://www.rapidtables.com/convert/color/rgb-to-hsv.html. Accessed 27 Sep 2019
12. Haralick, Robert, M., Linda, G.: Shapiro, Computer and Robot Vision, vol. I, Addison-Wesley, pp. 28–48 (1992)

13. Krizhevsky, A., Sutskever, I, Hinton, GE.: ImageNet Classification with Deep Convolutional Neural Networks. In: Advances in neural information processing systems. Available: https://papers.nips.cc/paper/4824-imagenet-classification-with-deep-convolutional-neural-networks.pdf. Accessed 27-Feb-2020
14. Liu, W., Anguelov, D., Erhan, DE., Szegedy, C., Reed, S., Fu, C.Y., Berg, A.C.: SSD: Single Shot Multibox Detector. In: European conference on computer vision, pp. 21–37 (2016)
15. Redmon, J., Divvala, S., Girshick, R., Farhadi, A.: You only look once: Unified, real-time object detection. In: Proceedings of the IEEE conference on computer vision and pattern recognition, pp. 779–788 (2016)
16. Ren, S., He, K., Girshick, R. Sun, J.: Faster R-CNN: Towards real-time object detection with region proposal networks. In: Advances in neural information processing systems, pp. 91–99 (2015)
17. He, K., Gkioxari, G., Dollár, P., Girshick, R.: Mask R-CNN. In: Proceedings of the 2017 IEEE International Conference on Computer Vision (ICCV), Venice, Italy, pp. 2980–2988, 22–29 Oct 2017
18. He, K., Zhang, X., Ren, S., Sun, J.: Deep residual learning for image recognition. In: Proceedings of the 2016 IEEE Conference on Computer Vision and Pattern Recognition (CVPR), Las Vegas, NV, USA, 27–30 June 2016, pp. 770–778 (2017)
19. Lin, T., Dollár, P., Girshick, R., He, K., Hariharan, B., Belongie, S.: Feature pyramid networks for object detection. In: Proceedings of the 2017 IEEE Conference on Computer Vision and Pattern Recognition (CVPR), Honolulu, HI, USA, pp. 936–944 (2017)
20. Microsoft COCO Dataset.: https://cocodataset.org/#home
21. NVIDIA TensorRT.: NVIDIA Developer, 24 Feb 2020. [Online]. Available: https://developer.nvidia.com/tensorrt. Accessed: 27-Feb-2020

Printed in the United States
By Bookmasters